GPS Tracking with Java EE Components

Challenges of Connected Cars

GPS Tracking with Java EE Components

Challenges of Connected Cars

By
Kristof Beiglböck

CRC Press
Taylor & Francis Group
Boca Raton London New York

CRC Press is an imprint of the
Taylor & Francis Group, an **informa** business

CRC Press
Taylor & Francis Group
6000 Broken Sound Parkway NW, Suite 300
Boca Raton, FL 33487-2742

© 2019 by Taylor & Francis Group, LLC
CRC Press is an imprint of Taylor & Francis Group, an Informa business

International Standard Book Number-13: 978-1-138-31382-8 (Hardback)
International Standard Book Number-13: 978-1-138-05494-3 (Paperback)

Library of Congress Cataloging-in-Publication Data

Names: Beiglböck, Kristof, author.
Title: GPS tracking with Java EE Components : challenges of connected cars / by Kristof Beiglböck.
Description: Boca Raton : Taylor & Francis, CRC Press, 2018. |
  Includes index.
Identifiers: LCCN 2018014010| ISBN 9781138054943 (pbk. : alk. paper) |
  ISBN 9781138313828 (hardback)
Subjects: LCSH: Vehicular ad hoc networks (Computer networks)--Data processing. |
Automatic tracking--Equipment and supplies. | Global positioning system. |
  Java (Computer program language)
Classification: LCC TE228.37 .B45 2018 | DDC 629.2/72--dc23
LC record available at https://lccn.loc.gov/2018014010

Visit the Taylor & Francis Web site at
http://www.taylorandfrancis.com

and the CRC Press Web site at
http://www.crcpress.com

Contents

Preface

The times of proprietary products dominating the software industry are over. Today more than 80% of all software applications are compiled of open source components and with the combination of Java and Open Source Software (OSS) alone there is practically nothing you can *not* program. The Apache Maven Build Tool has emerged to an automated project manager. The architect of a new application is in charge of composing the overall Project Object Model (POM). Once the project is defined each developer can create the software from scratch and begin to implement code with his domain knowledge.

Maven Central is hosting millions of artifacts and serving tens of million downloads per week, tens of billion per year, with growing demand. A Maven Server can easily be setup inside your company to support distributed software development for your team. Maven is machine automation of development, integration, testing, documentation and deployment. The infinite number of Java OSS Components is the modern challenge for software architects to compose an application.

Choosing the right OSS components already is a challenge and this book will demonstrate how to select Java and Java EE Components and compose the technical framework for your application before you start adding business code. Our business domain is GPS Tracking and we will see how to create individual components for a GPS Tracking System (GTS) and put them together to a running system. This process of picking JEE components and adding GTS features can be expressed in the pseudo equation

$$\text{JEE} + \text{GTS} = \text{JEETS}$$

where JEETS is pronounced like G.T.S., i.e. JEE T.S.

The book's index provides an overview over JEE and JEETS components. Instead of a Bibliography the book provides a 'Recommended Readings' Section of freely available Internet sources to be used while you are reading the JEETS code. In the recommended readings you can find book recommendations to explore each component in full depth – after you have gone through the book for a complete picture.

Instead of creating an application from scratch, which exceeds the limit and purpose of a book, we will work with the Open Source Tracking System Traccar. Traccar is a popular Java GTS that you can install out of the box and start tracking with your smartphone immediately. Traccar is a proven product, maintained very well and upgraded frequently by Anton Tananaev. Once Traccar is up and running we will take it as the running production system and put ourselves in the position to customize and replace GTS functionalities with new modules. This is a realistic situation for most professional developers.

We will analyze the Traccar GTS architecture to identify functional components like Device Communication Servers (DCS) and then create a JEETS DCS component that can run stand alone or be embedded inside an JEE Application Server. With this approach you will learn how to decompose a monolithic application into independant modules and then apply integration technologies to put the pieces back together with explicit endpoints to run them as one system.

At the end of the book we will have created a Maven Repository with well aligned JEETS components that you can use as seed components. Hopefully we will meet at jeets.org to create integration tests and improve the JEETS components according to your requirements. You will able able to create your own Java components and combine them to a new application or with your running production system. You should also become aware of the paradigm shift from large JEE application servers to micro services that can be scaled individually. You will be able to create an Enterprise Application running on an AS or you can create many components and assemble them to one server backend application or both.

The Self Driving Car (SDC) is the latest global challenge of steering a real car with software. You could not navigate a car through a city without information about the traffic situation. GPS Tracking has to be taken to a higher level to be able to track many cars at the same time and provide the aggregated traffic situation to each car. This constellation is referred to as Connected Car Technology and we will look at the impact on classical tracking systems.

CHAPTER 1

Introduction and Overview

CONTENTS

1.1   GPS TRACKING WITH JAVA EE

Vehicle and fleet tracking has become a large industry over the last decade and the market offers innumerable trackers for different appliances. With the smartphone a new tracker type was introduced as a multipurpose well performing *Client* computer with full connectivity to *any* sensor.

On the *Server* different GPS Tracking Systems (GTS) have evolved to track people, assets, vehicles ... *anything*. A vehicle tracking system can be used to observe a fleet and locate vehicles on a map in a browser frontend. The system can raise individual events to improve logistics, monitor fuel consumption, create travel reports etc. A transport company couldn't persist in the market today without a GTS and the software industry provides full fledged solutions to run a business.

Since the automotive industry has recognized tracking technologies as *the* live data source to actually direct cars through a complex environment the classical GTS architecture has to consider many new aspects. This book analyzes how the challenges of the Self Driving Car (SDC) exceed the limits of a classical GPS Tracking System. For a thorough analysis the widely accepted and well designed Open Source Traccar GTS [1] is reverse engineered in detail to dissect the major GTS constituents. Traccar is a monolithic application in Standard Java (JDK) running in a single JVM and we will re/model individual GTS components to reusable Java EE,i.e. JEE Modules.

The readers can witness the prototyping and modeling *process* and modify their own software. This book can help you to set up *your* tracking system by customizing the components. Every component is introduced in detail and includes a number of *design decisions* for development. The modules are combined into a customizable GTS – the JEE Tracking System JEETS . And they can be used as seed components to enrich existing systems including Java middleware and JEE application servers with live tracking. JEETS is a toolbox of GTS components combined to a bare bone level that you can also use in conjunction with Traccar.

1.2 THE CHALLENGE OF THE SELF DRIVING CAR

Navigation systems and digital maps have become extremely precise to localize a vehicle on a digital map with GPS coordinates. It was only a matter of time before tracking and mapping will be merged to actually guide the vehicle toward becoming a Self Driving Car (SDC). This challenge is currently changing the automotive industry from programming embedded software to hosting services and data crunching – in real time with really Big Data. The Internet has grown rapidly to serve people. Now it is being prepared to assist cars, sometimes within seconds, as they move.

The main goal in development of a self-driving car is to provide a calculated route and to actually control the car to follow it. At the first impression the SDC seems like a logical continuation of automotive developments. Existing intelligent Parking Assist Systems are noteworthy, since they actually take *complete control* over the car for a complex maneuver: heavy sensitive steering, driving forward and backward with high precision – *without human interaction*. This book will not explore the internal control of a vehicle but rather focuses on client server programming to provide useful information *fast*.

So the SDC is not really a revolutionary idea? Is it simply a normal evolution of technology? It is much more. The challenge is the live *coordination* of moving cars and the prediction of *coincidences* that could trigger events for cars approaching each other and avoid accidents.

The main difference of the SDC scenario to a classical GTS is that the latter only processes the tracking information of a single device! In the SDC scenario we need to evaluate submitted values of different vehicles and provide fast feedback to each vehicle as it moves. The system has to first evaluate and classify each vehicle message before comparing and aggregating different vehicles and finally supplying conclusions to each vehicle.

This SDC challenge requires a *remodeling* of a classical GTS to more independent components. We will combine these components on a higher level to form individual use cases which can be processed in parallel. In order to achieve this the JEETS will be modeled following some major guidelines:

Don't look into the past.
A classical GTS usually makes frequent use of database lookups to determine where the vehicle was previous to the submitted position.

Don't speculate on the future.
Some GTS keep up a connection dialog to wait for incoming events.

Don't wait for external resources.
The response time of an external resource is generally unpredictable. An external database represents a hardware resource and can easily become the main bottle neck of the complete system.

1.3   INTENDED AUDIENCE

Another main topic of this book is the reflection of JEE technology itself. Containers have diminished and can be created on the fly to execute a single method – and then be dropped again . . . JEE technology is slowly moving away from application servers to smaller independent Java (EE) components which are combined to a system. Many of these components will be introduced and applied for the creation of stand alone JEETS components.

In Java EE we usually speak of the architect and developer roles, which does not reflect the modern software development completely. The build tool and sophisticated test frameworks have become a core part of the development process. The Java EE platform is 'only' a higher level *specification* of many globally accepted API specifications related to each other, to the Java world and *the network*. Before actually starting an implementation the architect has to work as a software *composer* to create the built environment. All JEE client- and server modules developed in this book can be found in the GIT repository and can be compiled, tested, packaged and installed with Maven.

One typical problem to upgrade a standard Java GTS to a Java EE GTS is the direct connection to the hardware. Tracking messages arrive at the ports of a computer, a hardware. On the other hand in the JEE runtime environment, the application server implementation is designed for inversion of control (IoC) and the business logic should not include any code to interact with server, operating- nor file system. Otherwise scalability over different machines will not work!

Therefore this book can also help Java developers not only interested in GPS tracking, but in modern software design from many individual modules. The complete system design process is described bottom up beginning with the creation of data formats. The reader should be able to sense the fundamental importance of a persistence unit and a compatible network data format for data transmission, which actually pre/define the system and dictate each component's design. Since it is merely impossible to completely define data formats before creating a system we will describe how to create formats, which can be modified during the development and according to new requirements to an existing system.

Since SDC development in the automotive industry is highly competitive and therefore taking place under non disclosure we will focus on applying Java EE technology to existing GPS tracking systems. By providing live tracking information to Entity Java Beans (EJBs) a GTS developer can easily break the isolation of his GTS and combine it with other systems of the enterprise. Just imagine clicking on a truck icon on the map frontend to retrieve the actual truck load from the warehouse and other logistic systems.

1.4   SKILLS / SKILL LEVEL

This book does *not* introduce Java EE and you are expected to...

> install the JSE Traccar GTS
>
> install the Postgres database (with PostGIS)
>
> download / clone a GIT repository
>
> apply the build tool Maven (clean compile test package install ..)
>
> install the application server Wildfly
>
> install ActiveMQ
>
> configure the ActiveMQ resource adapter for WildFly

The book's source code and sample application are available to the reader on the book's website `jeets.org`. The book should be used as a hands-on instruction accompanying the source code. The reader is expected to download and compile the sources in a personal development environment. You can go through the appendixes all at once to setup your environment or you will be asked to setup required software by single appendixes as you read.

1.5   THE AUTHOR

The author is specialized on Geographic Data Processing and has worked for the automotive industry for more than a decade and witnessed the developments from GPS, digital maps, routing to navigation. The author has setup commercial GPS tracking systems with open source components and has been architect and developer in large automotive tracking projects. With the goal of a Self Driving Car GTS technologies were applied for real time tracking.

I

Data Formats and Relations

Message Exchange

CONTENTS

2.1 GPS PROTOCOLS

The new challenge of handling crowd sourced data flow of smartphones, cars and connected things is the reduction of data to streams without losing information. In the context of the SDC every single bit and every millisecond eventually define the information limit to support every car – for a single highly frequented traffic node – as well and fast as possible.

In the course of this book we will develop different JEE components especially for GPS tracking purposes. To start the development from scratch we will look at the basic information bits of tracking: GPS coordinates. GPS was thoroughly introduced in the first book and broken down to the following fields needed to describe a point in time and space:

longitude: $]-180°, 180°]$ decimal degrees east/west +/-

latitude: $[-90°, 90°]$ decimal degrees north/south +/-

altitude: meters above (+) or below (-) see level (0) [default]

timestamp: UTC[1] time and date

---

[1] Universal Time Coordinated, formerly Greenwich Mean Time (GMT)

event: The cause of sending this GPS coordinate

Tracking information, i.e. events should *always* be tagged with a coordinate GPS(lat,lon,alt,time) to describe *what* has happened *where* and *when*. For the time being the event stands for any (sensor) information to be transferred to describe the meaning or cause of a transmission. The submitted fields are transformed into Plain Old Java Objects (POJO) to become entities of a tracking system. (see Figure 2.1).

Let's look at these four fundamental GPS fields to gain an understanding of different data formats and how much space, i.e. bits, they require. As Java developers we want to handle the submitted values defined by Java primitives or -types as the server target format. Although a 32-bit representation in a float would be sufficient lat, lon and alt are commonly used as 64-bit double primitives. A vital question in a tracking context is how to transfer the GPS data over the network with a minimum number of bits without losing precision?

For the ease of use many trackers transfer information in a human readable text (ASCII csv) format:

```
keys         time            lon     lat     alt
values       ..,170312123644,120589,490234,0,..
formats      YYMMDDHHMMSS ddmmmm    ..
29 bytes     +---------+---------+---------
```

The first field represents a time stamp followed by lat and lon represented as digits and finally the altitude relative to sea level. This is useful for display purposes or to send tracking messages via SMS and many other human interactions. For an SDC scenario this format is a disaster.

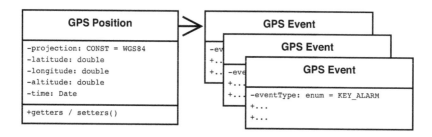

Figure 2.1   The server transforms the messages into related POJOs. GPS and events *can* be modeled in one or more system entities. Their relation *can* be modeled as 1:1 or 1:n as needed.

TABLE 2.1 Integer Coordinates with a Precision Field

longitude	dir	integer	precision	double
121.1234567	East	+1211234567	7	+121.1234567
121.1234567	West	-1211234567	7	-121.1234567
21.1234567	East	+211234567	7	+21.1234567
21.1234	East	+211234	4	+21.1234

In a text format at least one byte is needed for every digit (or in this case for every character) and separator which adds up to 29 bytes. Also it is unclear whether the altitude of zero is a default value or real (unlikely). The lat and lon values are restricted to 6 digits to fix their overall precision.

From the data processing perspective it's a good idea to use integers instead of decimals, if you compare FLOPS and MIPS performances of computer systems. The problem emerging from the SDC is the precision of lat and lon. Digital maps were used to provide an accuracy in meters and were subsequently refined for ADAS[2]. Since this is still insufficient for autonomous driving a new type of map is coming up with centimeter precision: the HD Map for machine reading.

On the other hand not every location is provided in centimeter precision. Therefore we could define lat and lon as integers and add a precision field (see Table 2.1). The precision value 4 represents the divider 10^4, i.e. multiplier 10^{-4} and allows defaulting to a precision or specifying it explicitly for any coordinate. The space needed for java variables is a 16-bit **short** for the integer, a 8-bit **byte** for the precision and a 64-bit **double** for the targeted variable in the server code. Now we have a 24 bit representation for lat and lon, which is much better than the text values listed above.

Depending on your application you might even consider submitting a small route, i.e. a collection of related position infos. This opens the possibility to tag each position with small deltas for lat, lon, alt. Yet this is a very project dedicated approach which can also introduce risks. In a networking context we don't have Java types and there is still room for compression by looking at the single bits. . .

2.1.1 Message Format Encoding

If you buy a tracker you usually don't have the chance to define a protocol and you must reverse engineer the one provided. Therefore we will look at a few tracking protocol formats and some typical implementation problems they pose. Afterwards we'll define our own JEETS data format with clear structures for fast prototyping.

---

[2] Advanced Driver Assistance System

We will now look at the tk102 (or tk103 etc.) as it is human readable (see [5]). A typical message String encoded by the tracker looks something like this:

```
008238008589B001141129A2302.7532N07232.2461E000
                        .0092142349.381000000AL000000F1
```

The server application receiving the message has to know the format in order to parse it into the final variables of your application:

```
008238008589   - trackerid / mobile number
B001           - alarm event
141129         - YYMMDD date
A              - valid data / V - invalid data
2302.7532      - latitude, format ddmm.mmmm
N              - N = north, S = south
07232.2461     - longitude, format (d)ddmm.mmmm
E              - E = east,  W = west
000.0          - speed k/m
092142         - UTC time HHMMSS
349.38         - altitude in meters (cm precision?)
1000000A       - bit representation of several events:
                 power on/off/external, ignition on/off, ...
L000000F1      - mileage
```

We won't go into the details, yet it's worth noting that lat, lon and time have the format of NMEA sentences and indicate that these values are propagated from the GPS processor output. Here is a Decoding Algorithm into a Java target format which should provide an impression of a Decoding Process:

```
2302.7284 = latitude, format is ddmm.mmmm
        N = north, S = south
>>> (d)dd + mm.mmmm. >>> 23 + 02.7284
>>> 02.7284/60 = 0.0454733333333333
>>> 23 + 0.0454733333333333 = 23.04547333333333
>>> Then multiply the result by -1 if the direction is S
>>> as N so 1*23.04547333333333 = 23.04547333333333
```

This sample conversion of the latitude indicates processing time, which again is something we should not waste for the SDC scenario. It may seem picky, but we want a precise look at transferred data to identify the bottlenecks. Another transformation could be required to calculate the *local time* from the UTC time and geocoordinates etc.

You should be aware of the altitude source. If altitude is important for your application you should not rely on the value supplied by the GPS unit! The height is determined from the satellites and highly depends on their constellation. Every time the GPS switches to other satellites the altitude precision increases or decreases depending on the geometry. The original NAVSTAR GPS from 1984 simply wasn't made for this. If you need to rely on accurate altitude values you should look for a tracker with a built in barometer. If you are building your own tracker you can get a barometer component with centimeter precision starting somewhere around 10 US dollars.

In the end every protocol is defined by groups (message definitions) of key-value pairs while the actual message is composed only of values. Due to data reduction client and server software have to agree on the (current) format of all sentences that define a protocol. The tk102 position message indicates some more challenges of data protocols.

If you take another look at the message string on page 10 it is obvious that the string length is vital information for decoding. If any value is missing, how could you decode the message? In the next chapter about TCP communication we will look at another GPS message string – this time with comma separated values (csv) which allows to submit empty fields as

```
"imei:359587010124900,,,13554900601,F,132909.397,,,"
```

A tracker specifies a certain protocol format for events triggered by external sensors or internal states like the battery level. The actual GPS processor protocol is usually the NMEA 103 format which was introduced for the very first GPS units around 1984. As we could see some trackers provide a setting to transfer NMEA sentences in a (complicated) raw format, like (d)ddmm.mmmm.

Later we will look at tracker architectures to understand how NMEA plays an important role inside the tracker controller to evaluate the GPS output and select high quality locations by taking accuracy, satellite constellation etc. into account. A tracker logic is responsible for selecting good events and formatting them into a tracker protocol format.

2.1.2   Message Type and ID

Sending a tracking message with a position and an event is the basis of tracking. Tracker protocols define different messages to submit event details. To set up a device communication software for a dedicated device you need to apply different value encoders and decoders repeatedly. Values for a certain key, i.e. latitude appears in different messages.

Commonly device communication software is designed for various messages of a tracker protocol and have many switch .. case constructs in their code to switch to message types and then field definitions. Many devices provide a header containing only message Id or place the Id as the first field in the message sentence. Many protocols tag the message Id with a $. As an example we'll look at some NMEA sentence formats

```
GLL - Geographic Latitude and Longitude:
$GPGLL,4916.45,N,12311.12,W,225444,A,*1D
GGA - essential fix data with 3D location and accuracy data:
$GPGGA,123519,4916.45,N,12311.12,W,1,08,0.9,545.4,M,46.9,M,,*47
```

to find that the messages $GPGLL and $GPGGA share some fields.

2.1.3 Message Catalogs

The `tk102` protocol format is helpful for a textbook and is sufficient to describe the process of encoding and decoding messages. For higher network traffic and smaller data sizes more sophisticated devices apply byte and bit manipulations for every data field.

To get an impression of a complex device you should have a look at Garmins Fleet Management Interface (FMI) [6]. FMI units provide a large number of features, i.e. messages used for fleet management solutions with navigation and messaging. Garmins FMI devices represent a connector for external truck sensors like a chat interface or a dash camera and much more.

The complete FMI functionality is specified by a large message catalog [7] where you will find message IDs for many different appliances:

A607 Waypoint Management and Driver Status

A611 Server to Client Long Text Message Protocol

A622 CAM 1.0 (Dash Camera Protocol)

The implementation details can be found in the Garmin Device Interface Specification [8] which has about 70 pages of technical details for encoding and decoding. We will not go into the specifics of extracting information from a decoded field and this specification can serve as a substitute for you to get a good impression of a complex device communication. The document describes data types, protocol layers and application protocols that you can study as needed.

Another good reason to look at the FMI devices in the SDC context is the fact that it is not a tracker! Although it does have a GPS unit and even more a digital map for navigation it does not hold a GSM unit. In order to communicate with a GTS or Fleet Management System the device has to be connected to a tracker with a GSM unit. The FMI device communicates with the tracker via physical protocols like the RS-232 or USB Standards. This is similar to the constellation of a tracker in a (self driving) car. The tracker is merely the device to submit messages and add a time and place stamp. We will look at some implementation details to deal with two devices a little later.

Now that we have a good idea of fields, messages and message catalogs we proceed to create a device communication server for dedicated message catalogs representing a device and its external sources or sensors.

2.2   TCP/IP COMMUNICATION

Generally we can assume that a tracking system is designed to receive live GPS information and telematics from a remote tracking device over the air (OTA). Since a modern tracking system should also be able to handle indoor tracking we don't want to restrict the system to OTA. The most general approach is to focus the Internet protocol TCP/IP and ignore the transport media – which is the actual benefit of TCP.

Transmission Control Protocol[3]

> The Transmission Control Protocol (TCP) is one of the main protocols of the Internet protocol suite. It originated in the initial network implementation in which it complemented the Internet protocol (IP). Therefore, the entire suite is commonly referred to as TCP/IP. TCP provides reliable, ordered, and error-checked delivery of a stream of octets between applications running on hosts communicating by an IP network.

For developers this information is concisely focusing on the implementation. We associate TCP communication with a port and an IP address on client- and on server side. And most important: an IP address always defines a hardware endpoint to software!

Port (computer networking)[4]

> In the internet protocol suite, a port is an *endpoint of communication in an operating system*. While the term is also used for hardware devices, in software it is a logical construct that identifies a specific process or a type of network service. A port is always associated with an IP address of a host and the protocol type of the communication, and thus completes the destination or origination address of a communication session. A port is identified for each address and protocol by a 16-bit number, commonly known as the port number.

For our software design we can simplify the complete communication by specifying `IP:port` combination at least for the server and, if available (i.e. online), for the client. You may miss the other common protocol UDP, which is also applied in various tracking devices. Anyway TCP/IP is a useful constellation we can work with and you can treat UDP implementations in a similar way. The data flow for the new system JEETS can be simplified to:

```
Tracking Device > TCP > Tracking Server > JeeTS
```

---

[3]From Wikipedia, the free encyclopedia
[4]From Wikipedia, the free encyclopedia

This is a pragmatic model to start an implementation and as we move on we'll see how to transfer data back to the Self Driving Car via JEE services. Note that the 'tracking server' is actually a Device communication Server (DCS). A DCS represents a single component of the tracking server and is coded to communicate with well known messages of the trackers Message Catalog.

By looking only at the TCP communication we can also cover the fact that indoor tracking is usually based on various different technologies, since GPS is not available. Besides indoor tracking we should keep a new emerging industry in mind, the Internet of Things (IoT), Industry 4.0 and so on. In the end we only need values for time, longitude and latitude for tracking.

For indoor tracking the altitude is sometimes used in a different way. Abstractly speaking buildings have one or more floors with a constant altitude and the altitude might store the floor number, which is used to load its ground plan. Check out the Minneapolis Mall of America in Google maps to get an idea of map matched ground plans.

2.2.1 TCP with Standard Java

Java was invented as *the* network- and Internet programming language and Java objects in general can communicate directly via TCP (or UDP) by design. Distributed systems, like the tracker and tracking server, need reliable communication. A Java server application can bind a socket to a specified port and exchange information via a point-to-point channel. During this session data can be exchanged in both directions.

Some tracking systems simply keep up the connection to a car as long as possible although most of the time the connection is idle and virtual. This model is usually driven by the GMS provider fees. For an existing connection the provider charges the actual bytes being transfered while every new connection requires additional data packets. It is most expensive to transfer every tracking message over a new connection with much more overhead than the actual information.

The Java developer does not have to deal with TCP layers when using the `java.net` package. While TCP is a generic protocol many more customized protocols like `http` and `ftp` are implemented on top of TCP and specified by the port in the URL. The Java `URL` class offers many methods to deal with these protocols to establish connections and exchange data. A simple position message could be transmitted via http URL encoding:

```
http://host:5055/?id=401258&lat=49.158&lon=12.864&timestamp=192412
```

The other important class to know is the `Socket`. After establishing a client server connection each side communicates to a local socket which can be associated with an instance, a session or channel of the relevant software logic. Sockets provide the ability to establish and distinguish multiple client server connections at the same time.

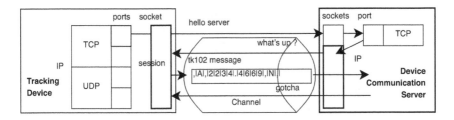

Figure 2.2 A tracker requests a connection to a DCS via IP:port. The server creates a new socket to establish a connection and receive the message/s. This can be achieved with the java.net package.

For a fast hands-on TCP lesson you can go through the Java Tutorial Trail: Custom Networking [9]. In the TCP section you can copy and paste the two classes EchoClient.java and EchoServer.java to let them communicate via the Echo Protocol [10].

We will use the EchoClient as a starting point for a tracker software. Although we want to create a tracking system in the long run it is *very* helpful to be able to send tracking messages to the server as needed in development time. Sending messages with real tracking devices can be very time consuming. In order to trigger a certain message you need to reveal the device to the sky. By placing it close to the window only half of the sky is visible. If you want to trigger a distance message you will have to walk the distance. Therefor we initially need some client software to send any message we chose.

Of course this can also be achieved with network tools like netcat. For automated testing unix scripts can be used to simulate high traffic with netcat commands. At this point we are interested in understanding the complete communication process and looking at it through our Java glasses.

The tk10x protocol introduced one page 10 is so popular since it defines complete messages in a single String. A developer can easily compose (encode) any String with any software and send it to a server counter part to de-compose (decode) it back to key-value pairs. This simplicity is used by many tracker vendors to add whatever they please. The format is not used very strictly and can easily become proprietary. Actually you will find many tracking protocols with ..102.. ..103.. and should carefully look at the actual messages coming from your hardware!

If you simply need to add basic tracker capabilities to an existing software then you can implement a simple tracker based on the EchoClient which can send a single string to the server like this:

```
String positionMsg = "imei:359587010124900,tracker,809231329,"
                   + "13554900601,F,132909.397,A,2234.4669,N,"
                   + "11354.3287,E,0.11,";
try ( Socket serverSocket = new Socket(hostname, port);
      PrintWriter out =
          new PrintWriter(serverSocket.getOutputStream(),true))
    {
      out.println(positionMsg);
    }
catch (UnknownHostException e) { .. }
catch (IOException e) { .. }
```

This is all you need to send a `tk102` message to a tracking system!

As this simplicity can cause problems on the server side most GTS should require tracking devices to be registered to the system before they can use its services. Another missing detail of this implementation is the lack of a server response. How do we know a GTS has accepted the data? The general solution is an Acknowledge message, or in tracking jargon: the `ACK`.

TCP is often introduced with the analogy to e-mailing. How do you know your mail with a meeting date has reached the recipient? You will only know if you get *any* response like a simple 'ok'. Although people would never reply "didn't get the message" a server can respond with "message corrupted" – a negative `ACK`, the `NAK`[5] Don't confuse the built-in TCP acknowledgment mechanism (behind the scenes) with the 'semantic' acknowledgment of your GPS data.

Especially when communicating over the air transmissions *will* be corrupted frequently. With many trackers you have the ACK mode as an optional setting and you should use it! Only after the server has acknowledged a message the client can dispose of it. Otherwise the message should be resent at the next opportunity. As a general rule of thumb "Communication without acknowledgment is no communication"!

Having said this we find another reason for the popularity of the tk102 protocol: Many servers do not send a reply, so for now we don't have to take it into account and the `BufferedReader` found in the `EchoClient` is not applied. With this strategy we can send a message to existing GTS without modifying it. We'll keep the ACK in mind and pick it up later for implementation.

What the listing does not show is the actual composition of the string – the content and its meaning. In a tracker hardware we have to deal with a number of predefined strings and if you search for 'GPRS data protocol.xls'[6] you will find a spread sheet with more than 40 sentences and server responses to get a picture. The protocol, being a catalog of messages, actually defines all tracker features and how to use them for your application.

We will send the tracking message to a real GTS later. But first let's look

---

[5]see [8]: 3.1.3 ACK/NAK Handshaking
[6]or any .. 102 .. 103 .. protocol

at the server side of the popular OpenGTS to understand the complexity of decoding tracking information.

2.2.2   OpenGTS DCS

It wouldn't make much sense trying to cover all protocol implementations in a book. We have only picked a simple protocol to show client encoding and server decoding in Java. Actually it is not always easy to obtain a protocol specification and even if you do you might not be authorized to use or publish it!

On the other hand a new customer can already have trackers installed and wants to move them to your GTS. Then you may need to reverse engineer a number of messages and fields without a spec. Like many development strategies protocol implementation requires Internet research. For most protocols you can find Java snippets to use and the following is an example of Java code for our tk10x protocol.

OpenGTS is a popular Java GTS you can download at opengts.org. At roaf.de/intro you can find an analysis of the GTS from architecture of the components down to the source code. OpenGTS has a very monolithic structure which makes it hard to extract isolated components. We will focus on its Device Communication Servers (DCS) responsible for decoding protocol messages on dedicated ports. Earlier we looked at the tk10x protocol as a readable example and we created a small tracking software to send tk10x message strings. At roaf.de/tracking you can study how the NMEA GPRMC sentence is decoded (with checksum) into target fields lat, lon, alt, time etc.

Let's look at the Device Communication Servers (DCSs) provided by OpenGTS[7]. In the original OpenGTS distribution you can run ant all with all additional components (Database, JDBC etc.) to create a DCS for the tk102 protocol. After running ant you should find a tk10x.jar in the ../build/lib folder which is simply started from the OS command line:

```
java <className> {options}
options:
[-tcp=<p>[,<p>]]  Server TCP port(s) to listen on
 -start           Start server on the specified port
```

For production releases you should use the provided batch or shell scripts runserver in the bin folder. Anyway the OpenGTS installation and compilation is not required to follow the tk10x message decoding – it is human readable.

For the following analysis you can create a regular Java project in your IDE and place the OpenGTS java sources in the src folder. This will not compile but does provide convenient code browsing. All DC

---

[7]We won't go into the details of the system nor install it.

servers reside in the `org.opengts.servers...` packages and we'll focus on `org.opengts.servers.tk10x` to identify the decoding.

The `tk10x` package holds four classes to form the DCS core. The `Constants` classname speaks for itself and you can find a mix of technical and tk constants needed to control the TCP connection. The `Main` class doesn't need much explanation and basically executes[8]:

```
TrackServer.startTrackServer(tcpPorts, udpPorts, cmdPort);
    new TrackServer(tcpPorts, udpPorts, cmdPort);
        ...
```

In this constructor you will find the class

```
org.opengts.util.ServerSocketThread extends Thread
```

which has almost 4000 lines of code and gives an impression of complex thread handling. The actual thread is started and configured with

```
private void _startTCP(int port)
```

where we find

```
sst.setClientPacketHandlerClass(TrackClientPacketHandler.class);
```

Whenever you are looking for a TCP decoder you will probably find a `Handler`, the main class to look for our 'business code' for `tk10x`, in this case a plain Java implementation. Let's look for our string message[9] and identify the decoding process.

Again we have a really big class with more than 2000 lines! The class combines different tk102 and tk103 messages which are parsed and decoded in the `parseInsertRecord_*(String)` methods. Fields are decoded in `_parse*` or `_get*` methods. Most remaining methods are used to manage the connection/s.

Figure 2.3 is showing a vital design detail in the context of the SDC. Many 'classical' tracking systems write the messages to and the actual GTS reads the messages from the database. Since a database management system is an external source with hardware access you can not rely on its performance or even availability. Therefore the messages for SDC should be propagated directly to the core system while the database access should be asynchronous. If the data volume for SDC is temporarily high a dynamic switch can apply filters to persist only every n-th message or even bypass the database completely.

---

[8]We are only interested in the TCP part, you can ignore UDP.
[9]search for 'imei:'

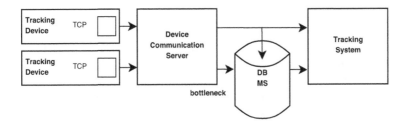

Figure 2.3 Some 'classical' tracking systems interact with the database (management system). In SDC scenarios the database access should be asynchronous.

2.3 JEE COMPONENTS

This book's title includes the vague and unofficial term 'Java EE Components'. What does this mean? What are they? So far we haven't looked at any of the JEE specifications at all.

The Java Standard Edition provides the core of every Java implementation. Nevertheless we have just seen how much effort is needed to implement a DCS with `java.net` and `java.io` packages. Would you rely on these thread and handler classes with more than 6000 lines of (historic) code in an SDC scenario where a tiny software bug might threaten lives?

JEE certified application servers probably have the largest global programming communities and the specifications keep all implementations compatible. On the other hand smart phones, websites, trackers and the Internet of Things have different requirements to network communication. Big data processing has lead to a de-composition of JEE application servers to be able to execute small pieces of code in various (virtual) environments.

In the next section we will look at our first JEE Component Netty.

Netty is a framework for high performance protocol servers and clients and has evolved from the JBoss community and their requirements. If we abstract the application server and other products to a collection of Java archives then it makes perfect sense to develop a single jar for networking requirements to be used by any application. Netty, as an example, provides Java networking classes for hundreds of client-server constellations.

This typical (self) organizational process can also be recognized in the Netty versioning history. While Netty 3 was developed in the JBoss community with `org.jboss.netty` package names, Netty 4 was detached from JBoss to become an product with `io.netty` packages. Take good care of the versioning in your built tool!

We will use the term 'JEE Components' for Java jars or libs that might not be specified by JEE while still being part of many implementations. A growing number of these components can be used in plain Java to become multi purpose tools.

Another term you can frequently read is 'Standard Java Component Technologies' which *is not* related to JEE although it can be applied there. JDBC is a well known example of such a component. Another example is `jcabi` [11]: "a collection of small and useful Java components, which are not big enough to make their own projects."

Modern Java development should begin with component composition with a build tool. Developers have to re-search for the latest components of their problem domain. In other words try to avoid writing technical implementations and focus on the business logic. The build tool takes care of resolving dependency conflicts with duplicate embedded subcomponents.

As we could see OpenGTS is based on standard Java with lengthy code and we will see how to drastically reduce the code by using the JEE component Netty.

2.3.1   The Netty Framework

So far we have created a simple tracker to send a message string with only a few lines of Standard Java. As long as you only need to send a GPS position to a server there is nothing wrong with this approach, no extra jar needed to blow up your application size.

Things get more complicated on the server side.

Every tracking hardware has its limits in the number of bytes per second and per port. For an SDC scenario technical limits have to be estimated and calculated down to a number of messages per second.

Imagine a scenario where a dedicated server application is responsible for analyzing a traffic junction. More generally speaking a 'traffic node' should refer to a place with problematic traffic flow[10]. The worst times, anyone knows, are the rush hours. You can find detailed traffic information for different days, daytimes, seasons etc. in the Internet or check Google Traffic for conditions in your town. With dedicated knowledge of the inner city and live data a bound box can be dynamically defined for a traffic node according to the hardware capacities.

With figures like these you can calculate the number of ports to use for a bound box or you can adjust the bound box to the number of messages per port. How much buffer does your hard- and software have to deal with high traffic? The scaling process can become very dynamic with live big data.

---

[10]Search for 'traffic mathematics' to begin calculating capacities, delays etc.

The Netty Framework [12]

> "Netty is an asynchronous event-driven network application framework for rapid development of maintainable high performance *protocol servers and clients*.
>
> Netty is a NIO client server framework which enables quick and easy development of network applications and greatly simplifies and streamlines network programming such as TCP and UDP socket server.
>
> Netty has been designed carefully with the experiences earned from the implementation of a lot of protocols such as FTP, SMTP, HTTP, and various binary and text-based legacy protocols. As a result, Netty has succeeded to find a way to achieve ease of development, performance, stability, and flexibility without a compromise."

As a starting point we have created a simple tk10x tracker earlier to send position messages from a remote client (we will actually send the message soon). We've used the plain Java TCP implementation of an Echo client and server in section 2.2.1 to get started. Please refer to the Netty Website and User Guide [13], a great single html page introduction where you can also find the Echo scenario. Let's look at the code differences of the TCP- and the Netty- Echo Client [14] and start by looking at the imported classes:

```
import io.netty.bootstrap.Bootstrap;
import io.netty.channel.ChannelFuture;
import io.netty.channel.ChannelInitializer;
import io.netty.channel.ChannelOption;
import io.netty.channel.ChannelPipeline;
import io.netty.channel.EventLoopGroup;
import io.netty.channel.nio.NioEventLoopGroup;
import io.netty.channel.socket.SocketChannel;
import io.netty.channel.socket.nio.NioSocketChannel;
import io.netty.handler.ssl.SslContext;
import io.netty.handler.ssl.SslContextBuilder;
import io.netty.handler.ssl.util.InsecureTrustManagerFactory;
```

Obviously no more `java.net` or `java.io` classes in the code! And we can identify the `io.netty.channel` and `io.netty.handler` packages as major workhorses of the framework.

```
EventLoopGroup group = new NioEventLoopGroup();
try {
    Bootstrap b = new Bootstrap();
    b.group(group)
     .channel(NioSocketChannel.class)
     .option(ChannelOption.TCP_NODELAY, true)
     .handler(new ChannelInitializer<SocketChannel>() {
        public void initChannel(SocketChannel ch) {
```

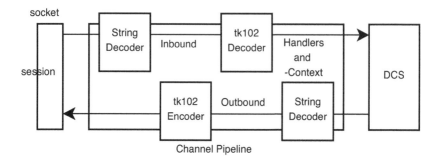

Figure 2.4 With Netty you can focus on implementing your customized en/decoders in Java combined with standard Netty en/decoders without worrying about sockets, session, threads etc.

```
        ChannelPipeline p = ch.pipeline();
        p.addLast(new EchoClientHandler());        }
    });
    ChannelFuture f = b.connect(HOST, PORT).sync();
    f.channel().closeFuture().sync();
} finally {
    group.shutdownGracefully();
}
```

The listing shows a typical Netty pattern and you don't really have to know the classes (see User Guide [13]). in detail in order to apply them. Basically you always create a `ChannelPipeline` and add one or more handlers for the incoming and upstream for the outgoing data. In this channel pipeline Netty hides all the details about sockets, threads etc. and takes care to connect and disconnect host and port. Netty also provides convenience methods like `shutdownGracefully()` that you should use in your code. Especially in network development time channels are opened and can't be closed until the code is working correct. This can be very disturbing as it can allocate folders and files which can prevent Maven from cleaning and even enforce a reboot!

A similar listing can be found in many client or server `main` methods to start Netty. The actual *business logic* can be found in the `EchoClientHandler` which simply extends a predefined Netty handler to provide method stubs. These methods will be invoked somewhere in the pipeline and sometime in a multithreaded event loop that handles I/O operations. Netty takes care of everything and provides simple Java constructs to place your custom code.

We will get to the client implementation when we build a software tracker in Chapter 8. For the time being our simple implementation is sufficient to send messages and if needed we could instantiate many clients to send messages asynchronously and randomly. Now we want to look at the server side to receive these messages from many clients.

The `EchoServer` looks similar to the `EchoClient` listing. While the client is constructed to transmit messages as it pleases, the server has to wait for messages all the time. This time we apply two event groups and use a Server bootstrap

```
// Configure the server.
EventLoopGroup   bossGroup = new NioEventLoopGroup(1);
EventLoopGroup workerGroup = new NioEventLoopGroup();
try {
    ServerBootstrap b = new ServerBootstrap();
    ...
```

In order to work with the `tk102` protocol the Netty pattern would look like this

```
protected void addSpecificHandlers(ChannelPipeline pipeline) {
    pipeline.addLast("frameDecoder",
        new CharacterDelimiterFrameDecoder(2048, "\r\n", "\n", ";"));
    pipeline.addLast("stringEncoder", new StringEncoder());
    pipeline.addLast("stringDecoder", new StringDecoder());
    pipeline.addLast("objectEncoder", new Tk102ProtocolEncoder());
    pipeline.addLast("objectDecoder",
        new Tk102ProtocolDecoder(Tk102Protocol.this));
```

which is also depicted in Figure 2.4. Note that String- and many more convenient en/decoders are provided by Netty and you can focus on implementing your messages in plain Java.

2.3.2   Device to Device Implementation

As we have seen a Garmin FMI device provides interfaces to many external sensors and internal features. Nevertheless it is not equipped with a GSM unit and needs to be connected to a standard tracking device via RS-232. Although every device provides dedicated information the tracker communicates with the DCS.

To implement the complete hardware chain you will have to deal with the trackers messages and then extract the Garmin messages which can have a completely different format. With Netty different message coders can be implemented in individual handlers to be consecutively aligned in the channel pipeline.

This section and the user guide can only describe how Netty supplies Java classes for hundreds of implementations and the Echo example shows a fully

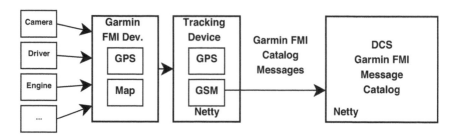

Figure 2.5   Although a Garmin FMI device provides a rich message catalog for internal Mapinformation and external sensors only the tracker device actually communicates with the DCS.

working network application build with Netty! We will create a tracker and a tracker server with Netty and you can add your own Netty process in between. The constellation of tracker, protocol and server is usually referred to as device communication and we will look at the device communication servers of the Traccar GTS in the next chapter.

Device Communication

CONTENTS

3.1 TRACCAR DCS

This book describes tracking technologies and provides tracking-, i.e. JEETS components as add-ons to existing applications. This book does not create a complete tracking system. By introducing Traccar as a complete and well structured tracking system we can analyze its architecture and use it as fallback for (yet) missing components. We will identify endpoints to place additional logic to any existing system. Experienced developers will be enabled to create a new system from Traccar and JEETS components in a complementary manner.

OpenGTS was the first complete Java Open Source GTS. The DCS modules are compiled to independent jars while the core system is built with the Ant tool and based on Servlet technology running on Tomcat. When Traccar came up as the second Java Open Source GTS many companies replaced the OpenGTS DCS with the better performing Traccar Device communication while sticking to the OpenGTS Web Frontend familiar to their customers.

We have introduced Netty as the TCP Framework of our choice. If you don't want to use Netty you can look at the alternative Apache MINA, a 'Multipurpose Infrastructure for Network Applications' [15]. The reason to introduce Netty was to now introduce Traccar's DCS architecture based on Netty. Instead of writing a DCS for the `tk10x` protocol we will look at Traccars decoding strategy.

Hands-on Instructions

In order to run, debug and analyze Traccar please follow the appendices 'Development Environment' and 'Install Traccar Sources' to set up your environment before continuing with the next section.

3.2   DEVICE COMMUNICATION SERVER

The core of GPS tracking is the device communication with many trackers via dedicated protocols and hardware ports. Usually for every tracking device type (and protocol) there is a Device Communication Server (DCS) to decode the protocol into the tracking system. For scaling and security purposes one single protocol can also be bound to different ports for different customers. Classical tracking systems store the information in the database where consecutive components can pick up this information (see Figure 2.3 on page 19).

The DCS is connected to a hardware to receive tracking messages (i.e. key-value pairs), parse them and propagate their information to the system core. A single DCS is usually bound to a port to receive a dedicated protocol and is a server software to receive live GPS information and telematics from a remote tracking device via TCP. The DCS processes these 'events' shortly after their occurrence.

In the next chapter we will look at the Traccar GTS and its DCS architecture. We will learn that every DCS of a single GTS creates the same Java *types* for a standardized data flow and process inside the actual tracking system. Especially for targeting the SDC scenario we should restrict the DCS definition to its core tasks:

listen to ports, create socket and session, receive tracking event

extract and convert data to the actual GTS model

The Device Communication Servers of OpenGTS are actually doing more:

add geographical labels, i.e. geomatch coordinates to address

administer tracking devices, create reports

create geozones for server side events

We will see later how to fulfill these tasks in a more modern way.

Note: The hands on instructions are described between solid black lines and the reader is expected to setup his environment in order to better follow the development.

Please go through the Appendix A 'Development Environment' to setup your environment and prepare your IDE to import the JEETS project sources.

Then you can proceed with Appendix B to download and 'Install Traccar Sources'.

This will take a while, but is highly recommended
to get the most out of the book and software as you read!

3.3   DC ARCHITECTURE

After describing the message flow from plain TCP to Netty we will now introduce Traccar's device communication Controllers for decoding more than one hundred tracking protocols!

In Traccar every device communication server is a Netty server analog to the `EchoServer` on page 23. Traccar is the main application managing the `ChannelPipelines`. Instead of building a complete application for your needs you should consider using Traccar. Add your protocol there and contribute it to Traccar – that's what Open Source is about.

To fast forward to our `tk10x` protocol you can prepare Traccar for the analysis. Please edit your configuration file and comment *all* ports except the `xexun` entry:

```
<entry key='xexun.port'>5006</entry>

<!-- PROTOCOL CONFIG
<entry key='gl100.port'>5003</entry>
            :
<entry key='cradlepoint.port'>5118</entry>
-->
```

Why `xexun`? As mentioned earlier there are many 'dialects' on the market and Traccar has implemented tk102, tk103, gps103 and `xexun`. The reason to pick xexun will be revealed in the following subsections.

With the above configuration Traccar is only serving the one protocol at port number 5006. Let's go inside Traccar and start in `Main.main`. The method gives an impression of the programming effort for a clean system with configuration, data, device, permission and connection managers and more.

We are interested in the creation of the `new ServerManager()`. The constructor searches the package `org.traccar.protocol` for implemented decoders. All protocols are derived from the `BaseProtocol` provided by Traccar for common handling. For each entry in the xml file a `new TrackerServer()` is started, in our case `XexunProtocol`. In the `TrackerServer` you will find the familiar Netty classes[1] customized for orchestration in the tracking system.

To implement your own protocol you can look at the two classes `XexunProtocol` and `XexunProtocolDecoder`. The Protocol class is setting up the Netty Pipeline which is executed *every* time a message arrives. The first handlers in the pipe prepare the String which is being parsed in the customized decoder:

```
XexunProtocolDecoder extends BaseProtocolDecoder
```

The actual parsing takes place in the method `.decode` and the first lines serve the message or sentence

```
Parser parser = new Parser(pattern, (String) msg);
```

Traccar generally uses predefined parsers to chop sentences into keys and values. By looking at the patterns you can easily tell their field sequences.

```
private static final Pattern PATTERN_BASIC = new PatternBuilder()
    .expression("G[PN]RMC,")                // Message ID
    .number("(?:(dd)(dd)(dd))?.(d+),")      // time
    .expression("([AV]),")                  // validity
    .number("(d*?)(d?d.d+),([NS]),")        // latitude
    .number("(d*?)(d?d.d+),([EW])?,")       // longitude
    .number("(d+.?d*),")                    // speed
    .number("(d+.?d*)?,")                   // course
    .number("(?:(dd)(dd)(dd))?,")           // date
        :
```

What you need to know when adding a protocol to Traccar is that *in the end every* decoder is creating a `org.traccar.model.Position` object as output. This object defines the internal Traccar format and is aligned with the database. We will look at this entity and entity management in the next section 'Data Modeling' very thoroughly.

For now let's get ready to send our first message by installing a tracker on your smart phone.

3.3.1 Traccar Client/s

The era of smartphones has equipped every owner with a very powerful tracking device. For developers it is not rocket science to make use of some kind of 'location API' to retrieve relevant values from the OS.

Besides a tracking server Traccar provides a tracking app for Android and

---

[1]Note that Traccar 3 is based on Netty 3 with `org.jboss.netty` packages!

IOS. The installation is simple and you should manage to install it to get ready to collect tracks. Don't forget to activate the port in xml. As soon as we start building the JEETS tracker the collected tracks will serve debugging and development.

3.3.2 myLiveTracker

The project `myLiveTracker` written by Michael Skerwiderski[2] is a very popular Android tracker for GTS (and DCS) development as it can emulate different protocols. The reason we have configured the 'xexun' protocol and port is that we will use `myLiveTracker` to emulate Xexun tk102.

Please install `myLiveTracker` on your smartphone.
Choose Settings > Protocol > Tk102 Emulator,
Close connection after transmission
End message with end of line
Determine your external IP - you might have a dynamic IP
which changes every time you log off and on again!
You may have to add forwarding of port 5006
to your firewall and or router.

3.3.3 GPS Test App

Before starting `myLiveTracker` you might want to install another useful helper: GPS Test by Chartcross Limited[3]. This App visualizes the satellites of six GPS systems NAVSTAR, GLONASS, GALILEO, SBAS, BEIDOU, QZSS and provides a lot of useful information. The most important information is the 'first fix time' which indicates how long it took to determine the first location. If you are indoors you can look at the satellite constellation and see how your device can only work with the satellites visible through a window. Once you have a fix you can switch to `myLiveTracker` and start tracking.

3.3.4 Debugging a tracking message

In this book we will also develop a JEETS tracker application which will be even more convenient than using your smartphone. For the time being we will use an app to send the first tracking message.

Now you should have setup Traccar and your environment with port 5006 for xexun from `myLiveTracker` and/or port 5055 for the Traccar client

---

[2]play.google.com/store/apps/details?id=de.msk.mylivetracker.client.android
[3]play.google.com/store/apps/details?id=com.chartcross.gpstest

OsmAndProtocol. Please pick your decoder in the org.traccar.protocol package.

We will use the XexunProtocolDecoder for debugging. Open it in your IDE and place a linebreak point at this line:

```
Parser parser = new Parser(pattern, (String) msg);
```

Next you can start Traccar *in debug mode* and then start your smartphone tracker. As soon as you see the IDE highlighting the linebreak point you can turn off the tracker again and step through the Traccar code...

In the debugger you will find the (String) msg from mylifetracker:

```
GPRMC,104641.2,A,4901.8459,N,01206.1216,E,0.0,0.00,
                    160117,0.0,E,A*3C,F,imei:356425050313211,
```

which is similar to our tk102 message. If you look at the pattern in the Traccar decoder source you can easily find the message fields with its regular expression for parsing.

You can go through the same procedure for any Traccar protocol with an actual tracker hardware. In the Traccar log you will also find important (session) information:

```
INFO:  [838043F7] connected
DEBUG: [838043F7: 5006 < 87.160.25.49] HEX: 31373031313631303436353
           4d432c3130343634312e3 ... 2c454142443334352c303342460d0a
INFO:  [838043F7] id: 356425050313211, time: 2017-01-16 11:46:41,
              lat: 49.03077, lon: 12.10203, speed: 0.0, course: 0.0
INFO:  [838043F7] disconnected
```

The HEX representation is the raw encoded TCP information before it is transformed to a string, while the ascii (Latin) listing represents part of the decoded information in a org.traccar.model.Position entity. This entity is persisted in the database and we can find the record:

```
366;"xexun";6;"2017-01-16 11:53:57.958";"2017-01-16 11:46:41.002";
"2017-01-16 11:46:41.002";TRUE;49.030765;12.1020266666667;0;0;0;"";
"{"signal":"F","ip":"87.160.25.49",
   "distance":89.62,"totalDistance":2246.76}"
```

with some empty "" or 0 values and a set of attributes {..} that could not be matched to existing Java fields. Please look into the SQL creation script of the positions table to find the database types of the above record. The position entity and positions table will be our starting point for an effective modeling of your domain. As we will see the Relational Java Object Model (ORM) and Databases Entity Relation Model (ERM) is much more than a message with fields (or sentence with words).

Relational modeling should reflect the complete system design.

3.4   CONCLUSION

At this point we have analyzed device communication with the most simple message format – a single string with all values. The idea was to go through the process of extracting a message and transforming it to a system entity. If you have expected more support to decode your tracker's protocol there are simply too many protocols and types on the market and most of them are proprietary. Anyway the `org.traccar.protocol` package should provide enough material to figure out how to decode your protocol. You can also ask for support at `traccar.org` and `jeets.de`

For an SDC project a client software is mandatory and we will develop a client for any Java runtime environment on the end device. If we address Android as a standard environment for the majority of user devices we can rely on a Java VM to run our client as the tracker. In other words JEETS is build on a JEETS tracker and its counter part the JEETS DC server to wrap the complete TCP stuff with Netty. Next we will define a general purpose protocol definition.

Data Modeling

CONTENTS

4.1 DATA MODEL

In complex Java EE software running on application servers data is exchanged in Java types, objects and references. These system entities should be limited and well defined for all components. In times of distributed systems we should not even rely on a specific programming language and rather focus on protocol specifications.

The first section about device communication outlines how well defined messages with ordered values are transferred over the network. In order to process the data on the server side each value has to be related to a Java Type key common to sender and receiver according to a protocol.

A complete protocol is composed of a number of message types describing a complete client-server dialog with all available features. More generally a scientific or diplomatic protocol describes a complete process with all inter-mediate events from beginning to end. This was indicated with the Garmin FMI message Catalog in Section 2.1.3. In such a case a protocol becomes mandatory and the application has to manage intermediate protocol states to follow the order of predefined events. When you implement a new tracker usually only a part of the provided protocol messages are actually transformed in a DC server according to the application needs.

Technically a tracker message is just a collection of variables. Semantically it describes some kind of event which has occurred during tracking to trigger the submission. A device communication server's task is to decode the submitted bits and bytes into some Plain Old Java Object (POJO).

Especially in tracking business with restricted bandwidth and connectivity the message size should be minimized for a safe transfer in little time. A regular GPS tracker can *not* generally be associated with a client software as a part of a distributed application. A tracker can be any (hardware) device sending serialized values from various sensors and no language specific objects. A tracking system can have many purposes and may reject many tracker messages in the first place. On the other hand an existing GTS can grow due to new requirements.

As an experienced developer you know that every application requires a model. In general a model is a collection of *all* data pieces (fields) grouped in entities, which are related to each other. Any good system design is based on the integrity of the data model. The model describes data formats and their relations in the (tracking) application. We will analyze how a good model can support elegant development of well performing processes.

4.2   MESSAGE TO ENTITY

A DCS is the entry point to a GTS and as we are focusing on Java the first task is to create Java objects from incoming messages. In the case of Traccar we have seen that *every* protocol decoder is creating a `org.traccar.model.Position`[1] as the return value of its `.decode(..)` method.

Traccar's `Position` object is pretty straight forward and we can easily find the base coordinates `GPS(lat,lon,alt,time)` we have defined in Section 2.1. There are a few more fields to enrich the information for the GTS, like GPS- and servertime to determine the network latency[2]. If you look at the `Position` source you can find a large number of constants compared to the limited number of fields:

```
public static final String KEY_SATELLITES = "sat";
public static final String KEY_EVENT = "event";
public static final String KEY_ALARM = "alarm";
public static final String KEY_STATUS = "status";
public static final String KEY_ODOMETER = "odometer"; // meters
public static final String KEY_BATTERY = "battery";
public static final String KEY_RFID = "rfid";
public static final String KEY_IGNITION = "ignition";
    ... etc.
```

---

[1] For OpenGTS `org.opengts.servers.GPSEvent` indicates the standard target entity.
[2] To ensure synchronized clocks the server should also use a GPS time source!

These constants are usually 'collected' during the implementation of additional protocols to reflect a certain type of tracker and event. For example the ignition key only makes sense, if the tracker is connected to a vehicles ignition signal in some way. For person trackers this flag makes no sense. The tracking system can and should use these constants to dispatch the incoming positions to different applications.

In the end Traccars position object represents a basic GPS message like the good old record set, the historical predecessor of a software object. Semantically *a* position is *one entity* of the complete system. Generally any application deals with a well defined number of entities to model the problem domain – with data structures. By relating entities to each other, like a `Person` to a `Device` with `Positions`, we can build a complete *Entity Relation Model* (ERM) as the data-base (or base data) for the application.

4.3   RELATIONAL MODELS

While an ERM is usually associated with a database model and table/row/s as entities, it is important to perceive it as general purpose relational data for *any* system development. The process of data modeling forms the basis for a system and its quality has vital impact on *every* implementation. Actually data design is the first and fundamental programming step for a system.

Another way for relational modeling is the *Object Relational Model* (ORM). For example you can also create a relational model with JavaScript Object Notation (JSON) without using any database technology. Technically there are countless ways to model an application domain. The core is always a semantic relational data model which is expressed in the nouns and verbs of the customer requirements.

With object/relational (ORM/ERM) mapping objects and database records (rows) can both be used in a synchronous way by applying the Java Persistence Architecture (JPA). The SDC context requires fast processing of incoming events and the software architect should take good care of which model to use when and how. As the ORM lives in the restricted RAM and the ERM is persisted on unrestricted hardware there are significant performance differences. Database queries can always be held up by other concurrent resources.

Let's explore the Traccar OR model represented by Java classes in the `org.traccar.model` package. In the last section we have looked at Traccar's `Position` as the output of *all* DC servers:

```
public class Position extends Message extends Extensible
{
                                      long id
                      long deviceId
      double latitude
      double longitude
      Date    deviceTime
         :        :
```

We can only transform a GPS message to a Java entity, if at least `lat`, `lon` and `timestamp` are available (required fields). On the server side with a lot of positions an identifier is needed for distinction and therefore the position has an `id` field which you will find in the base class `Extensible`. With a little database experience you know that database systems (DBMS) generate identifiers while persisting an entity.

What, if we want to use the entity without database and without ID?

One of the biggest achievements of the Object/Relational Mapping frameworks is that the developer does not need to know nor touch any IDs or ID relations (i.e. primary and foreign keys). As a rule of thumb you *should not* code any logic based on IDs. In Section 6.5.1 we will see how the entity manager takes over this task.

The next mandatory value of a tracking message is some `deviceId`, which can be the IMEI, phone number... whatever the system requires – and this one is not generated by the database. It only has to be unique inside the model like a primary key of a table row. We can peek into the Traccar database[3] (ERM) table `positions`:

```
CONSTRAINT pk_positions PRIMARY KEY (id)
```

What makes the `Position` class relational is the parent class `Message` which introduces the `deviceId`.

```
Position extends Message extends Extensible
```

In the database you can locate this relation in the table definition:

```
CONSTRAINT fk_position_deviceid FOREIGN KEY (deviceid)
    REFERENCES public.devices (id) MATCH SIMPLE
```

This relation has an impact on the GTS functionality: if you have successfully recorded a tracking position as described in Section 3.3.4 remember that you had to register your device before the GTS accepted its messages. In other words the message (i.e. position) can only be persisted if the devices table holds the `deviceId`.

We can conclude that an entity has values (`lat,lon,time...`), an identifier (`id`) with relations (`deviceId`) to other entities of *the* model. The DC server extracts values from messages and transforms them to entities. A simple message *can require* a (database and/or object) relation in order to be accepted by the GTS (or actually by the model). In the next chapter we will use this knowledge to look at message relations and how you can model these relations into your protocol.

---

[3]For this book we will use PostgreSQL [16]

4.4 THE TRACCAR MODEL

As we are focusing on Traccar's *Relational Data Model* we can use Traccar code and the database to identify all fields of the systems ER- and OR model. One of the main entities for any tracking system is the `Position`:

```
public class Position        CREATE TABLE public.positions
{                            (
  id integer NOT NULL DEFAULT nextval..,
  private double latitude;     latitude  double precision NOT NULL,
  private double longitude;    longitude double precision NOT NULL,
  private double altitude;     altitude  double precision NOT NULL,
  private Date deviceTime;     devicetime timestamp wo tz NOT NULL,
    :                            :
```

One important thing to note is the *single* position entity and the table for *many* positions. While a database table usually stores any number of records an entity always refers to a single record (or row) represented in a single object. The listing also shows the `NOT NULL` constraints to prevent storing incomplete positions. Anyhow constraints can also be *annotated* in Java and we will look at the persistence model later. Let's look at some fields and their implications.

ID

The `id` in the database table should be provided automatically by the database system (DBMS) with a sequence generator to ensure relations via Primary Key. An `id` provides distinct identification of records (entities) which is vital to relate entities to each other. On the other hand the client software *should not* supply `id`s and leave it to the database engine to generate them as needed.

Device Time

Earlier it was stated that a message definition consists of keys with well defined (language neutral) data types. Nevertheless a Java `Date` object is nothing but a wrapper for a `long` value and can be mapped directly.

Device ID

The `deviceid` is a 'speaking' identifier and 'belongs' to the device and should be provided by it. The IMEI[4] can be perceived as an identifier hardwired to the device. If the client software accesses this number in a safe way and adds it to the protocol messages you can rely on it. And you gain implicit validations of the GSM network, since it uses the IMEI to identify valid devices.

Sometimes there might also be a reason to choose the SIM cards IMSI[5]

---

[4]International Mobile Equipment Identity
[5]International Mobile Subscriber Identity

as the identifier. This is helpful when you evaluate different trackers with the same vehicle as it makes it easier to change the tracker type (in the prototyping phase) by simply exchanging the SIM. We will not use the SIM as identifier.

Whatever you choose it is recommended to use some kind of 'official' identifier instead of generating them in the database! In the Traccar GTS the identifier can be any *unique* character string.

Validity

Generally tracking systems should have something like a validation module to sort out positions invalid for the context of an application. For example when a car stops at a place with poor satellite visibility the position can instantly jump more than a kilometer before it finds a good constellation for a fix again. These positions can cause problems like jumping out of a geozone to trigger wrong entry and exit events. Therefore validation and plausibility checks belong to a reliable client and server system.

Character Strings

In general machine to machine communication does not *require* strings at all and architects should try to avoid them for device communication. The above mentioned `deviceid` for Traccar is an exception and provides the freedom to define your own id like `MyAndroidTracker`. Strings are only needed for human readability. You should also be aware that it *does* make a difference what kind of characters you put in a string. You might successfully develop a network protocol with an `ASCII` or Latin character string as a place holder. Later you run into problems on the production machine which also uses `UTF-8` Strings. Before adding a string to your protocol you should think about alternatives like using the IMEI as a number etc.

Let's look at two string fields in the Traccar position:

Address

In Traccar this string field is 'reserved' for reverse geocoding on the server side. A reverse geocoding service derives an (preconfigured) address[6] from the lat and lon coordinates. As reverse geocoding can be an expensive process it is usually delegated to an isolated service. You can choose a reverse geocoder from a number of suppliers and pay for the service by the number of geocodings per time unit. Or you can setup your own service which has the advantage to add your customer's places like 'headquarters', 'warehouse north' etc. which are familiar to all employees and drivers. Of course you could also fill this field on the client side to indicate `checkpoint#3763`. Anyway we will associate this field to the Traccar server and will not use it on the client side.

---

[6]From the Administrative Hierarchy of the Map

Protocol

This is another string of the Traccar position to be stored in the database. For the sake of simplicity we will only apply one protocol per port and this field is not used to determine a port from the protocol name. It does not make sense to fill this field on the client side.

Message ID

This field is not explicitly defined in the Traccar model but can be found in many protocols to support the network transfer. When the client receives an ACK response it can clean up the sent messages and by adding a `messageId` to the response the client logic can clean up the successfully transferred messages. Many trackers automatically resend a message after a certain number of seconds, if they don't receive an acknowledgment.

Note that the `messageId` is *not* mandatory for a single Netty session processing a single message. The point is that you can add fields to the 'network model' that will not end up in the system model.

CHAPTER 5

Object Relations

CONTENTS

5.1 GOOGLE PROTOCOL BUFFERS

IF – and that's a big if – you have the chance to design and implement a protocol on the client side we introduce the most effective way to create and modify (i.e. prototype) a protocol during development time and serialize the message for the network transfer.

If there are so many protocols out there why not figure out a standard way to implement *any* protocol? The premises is simple: Serialize a number of (ordered) typed values and de-serialize them to their keys, in our case Java typed variables. Protocol messages are actually exchanged with any client since any kind of information can be called protocol and a client is a client, because it works with external data. Google has addressed the lack of a standard protocol technology and came up with

Protocol Buffers [17]

> "Protocol buffers are Google's language-neutral, platform-neutral, extensible mechanism for serializing structured data - think XML, but smaller, faster, and simpler. You define how you want your data to be structured once, then you can use special generated source code to easily write and read your structured data to and from a variety of data streams."

Please to go through the developer Guide [17] to get a quick intro with a simple message describing a person. Just think of our `Position` when you read `Person` to get started. Please use this introduction to install the protobuffer compiler `protoc` on your system. You can then modify Google's example to compile the snippets of the following sections.

If you have already worked with JavaScript objects you can think JSON to get familiar with protobuffers. The simplicity of using protobuffers lies in the `*.proto` file, a simple text file to define messages with fields. From the previous sections field evaluation we can conclude that the system and network models basically map to each other while they are not identical. We will deal with the implementation of our protocol with protobuffers in Chapters 6 and 7 and we will deal with this asymmetry. In this section we will focus on the design of the proto file to define the network model.

After we have looked at the positions table (ERM) and the Position entity (ORM) let's create the position with protobuffers:

```
message Position {            |   postges types
    int32  deviceid   = 1;    |   integer NOT NULL
    uint64 servertime = 2;    |   timestamp
         :                    |        without time zone NOT NULL
    // decimal degrees +/- N/S |
    double latitude   = 6;    |   double precision NOT NULL
    // decimal degrees +/- E/W |
    double longitude = 7;     |   double precision NOT NULL
         :                    |        :
```

The protobuffer message is on the left side while the right side shows the targeted Postgres types. In general protobuffers are language neutral and therefore have to be mapped to the receiving environment. The listing is skipping the Java types as we will encounter them soon.

If you look at the complete message in the `traccar.proto` file[1] you will note the missing `id`, `address`, `protocol` fields – as discussed above. A useful measure is to supply important remarks for developers by adding comments to the message and fields. They will be processed to `javadoc` comments where they can be very helpful and time saving when hovering over the variables.

When you create the initial proto file you should be aware of the explicit ordering with '= 2' expressions. This looks unnecessary since the fields appear in exactly this order. The reason is the forward- and backward compatibility of a protocol over time. If you recall our first `tk102` message with empty fields it should become clear that the protobuffer internally applies the order to define voids, missing values.

In earlier versions there used to be a `required` token for fields and you can find interesting discussions about them, which are quite educational for any software developer. All you need to know for now is to order your fields by

---

[1]see JEETS Repositiory: /jeets-protocols/protobuffers/traccar.proto

importance. lat and lon are *required* for tracking and should keep their ordinal number forever, while you can add fields at the end to support additional (project related) logic.

5.2 BULK MESSAGES

Defining a message in a proto file is as simple as can be. Anyway there are many design implications you should be aware of. The main task is to model the messages towards the systems relational data model. Creating a position message is a good start for prototyping since it can be used to setup the implementation and development cycle. The position message can be sent to Traccar to result in a position entity which is then stored in the positions table.

Anyway our first message does not justify the use of protobuffers. There is much more you can do with them to break out of the 'classical' parsing of isolated messages. Many trackers have some kind of internal database to store GPS fixes immediately and regardless of an existing connection.

Especially SDC scenarios can require the serving of GPS coordinates after a precise pattern. For a simple example every 10.00 seconds or (speed dependent) precisely every 50.00 meters. A sophisticated client software can constantly analyze the GPS unit's (internal) calculations with DGPS enhancement etc. to interpolate the parameters and serve the best possible coordinate for a given time or place. With an internal client 'database' the coordinates can be *buffered* internally to balance the unreliability of the GSM connectivity. Note that Traccar's position has three timestamps fix-, device- and finally servertime to document the positions creation history.

An internal client logic can deal with the buffered messages to send them when there is a good connection and it can leverage the number of messages to send at one time. More than that the client can constantly traverse the buffered coordinates to align and optimize them in some way. In any case a maximum number of bytes should be defined to prevent blocking a server port for too long.

Many classical trackers offer a *bulk message* to transfer the internal database in one message. Besides implementing additional message formats with new parsers this usually does not provide a control mechanism for the total message size.

With proto buffers it is very simple to transfer any number of positions by simply adding another message to the proto file:

```
message Positions {
    repeated Position position = 1;
}
```

That's all. Now we can implement and configure one port to receive `Positions` and the client can set the number of messages as needed. And of course it can implicitly be used to send a single `Position` – wrapped in a collection as a tiny overhead.

If we agree to use the `Positions` protocol instead of `Position` we can already spot the next opportunity to reduce the size. As every `Position` from a single device holds the same `deviceId` we can reformat the message to:

```
message Positions {
    int32  deviceid = 1;
    repeated Position position = 2;
}
```

and remove the `deviceId` from the single position message. In other words the `repeated` keyword is used to create collections of messages – without using explicit IDs. And it can be used as a *buffer* to collect positions on the client side before it is converted into the network format for streaming with a single method call.

5.3   OBJECT RELATIONS

We are still in the process of relational data modeling after Traccar's data model. The initial ideas leading to object oriented software development were driven by language analysis. A noun was modeled into an object and objects are related by verbs, i.e. methods [19]. When we go back to the Traccar model we can find the relation device to position. Positions are always created by a device. Therefore we can rename the `Positions` message to a `Device` message:

```
message Device {
    int32  deviceid = 1;
    repeated Position position = 2;
}
```

to model the `1:n` relation of one device to many position entities. Again this is more than simply changing the name of the message and opens new modeling options.

One of the commonly used sensor data is the devices battery level and is part of many protocol messages. With protobuffers you could relate this piece of sensor data to the device:

```
message Device {
    int32  deviceid = 1;
    repeated Position position = 2;
    int32  batteryLevel = 3;
}
```

which adheres to common sense as a position does not have a battery level. With protobuffers this modification can be applied for an existing software since we have added the new field at the end without modifying the existing order. Previous client implementations (out in the field) are still working without being aware of the battery level!

5.3.1 Model Implications

In the modeling process you should always be aware of the semantic implications. For vehicle tracking the battery level is marginal since the tracker is fed by the vehicle's battery which is constantly being loaded. When tracking people with disabilities in ambient assisting frameworks the battery level can become *vital* !

A typical person tracker depends on its internal energy source which dominantly defines the devices size. By rule of thumb a rechargeable battery with 1 *Ah* (or 1.000 *mAh*) can track for about 10 hours highly depending on the persons activities. The standard requirement is to cover one day and reload the device when the person goes to bed. There is no way to determine the exact timespan depending upon the number of connections, the satellite visibility etc. varying every single day. Anyway the tracker – and *not* the tracking system – should report a low battery event to raise a notification at the emergency call center.

If you think about our `Device` message in this context you can identify a dangerous pitfall. Let's say the device message reports 17 percent battery power and a low battery event is planned below 15 percent. The problem occurs if the battery power becomes too low to establish a GSM connection and prevents sending anything, not even a low battery event.

Therefore a low battery event should always be generated in the client that should be smart enough to send the event at 17 percent even if the rule is 'below 15'! In the described scenario the server could never trigger a low battery event by the rule: $< 15\%$. A thorough server implementation should generate some warning anyway.

Most message modifications should be implemented thoroughly and the associating the battery level to the device makes sense. The implementor of the software logic should make sure that the battery level of the *device* is updated for every new *position* added. Only then can you conclude that the battery level represents the value at the timestamp of the last position.

If you are interested in observing the battery consumption in detail – move the battery field back to the position message. Modeling is programming and models imply – and dictate – system wide rules.

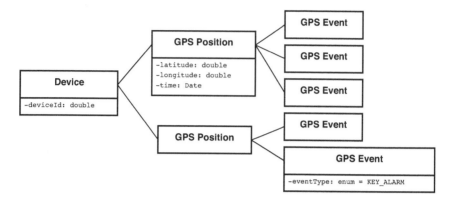

Figure 5.1 This single device message holds an object relational model with one device-, two position- and five event entities.

5.4 ORM AND ERM

We have seen how to model relational messages with protobuffers. Now what exactly is an Object Relational Model (ORM) and how does it relate to an Entity Relational Model (ERM)? Trying to look it up can become even more confusing and the terms are applied differently in various contexts. Let's not rely on any theoretical differentiation and rather look at the practical implications.

If you are familiar with the Unified Modeling Language (UML) you are aware of the difference between a class model and an object model. Objects are created, instantiated after the blueprint of a class definition. In database models, often imprecisely referred as ERM, a table ($\sim$ class) defines the valid content by rows with datatypes while each row ($\sim$ object) holds a concrete content – a semantic entity of the system domain.

The single string message was introduced on page 10 as an example of one protocol sentence with a well defined and limited number of fields. The message size allows estimates about the maximum server performance to determine the number of ports needed for a certain scenario. The protobuffer device message is relational and therefore its maximum size is *not* fixed.

In our context we will relate *one* device message to *one* ORM. The point is that the message definition, similar to a class, does *not* define how many positions are attached to the device. One device message including five position messages adds up to six instantiated objects, i.e. messages with individual values. Therefore this message represents *one* distinct ORM with *six* concrete objects (see Figure 5.1).

Later we will create a Persistence Unit (PU) to represent the complete model with Java classes. Then the device message being one ORM will be *transformed* into one device entity with five related position entities. This new ORM based on the PU can be passed to any application which can easily navigate through the model to pick any value required for the application logic. The navigability is one of the main advantages of object relations.

By creating relational messages we are defining many different ORMs which represent a certain detail of the systems data model. These messages usually do not (and should not) represent the complete ERM but rather a relevant fraction to model a use case of the system.

In theory you could create the network model as an image of the complete ERM and send the relational data in a single message and a single protocol. The complete database! This is generally the wrong approach and the system architect should design every message, i.e. ORM package for dedicated use cases with limited data size.

5.5 DESIGNING ORMS

For a productive prototyping phase it is very helpful to define a 'root' message which can imply related messages. The advantage is that we can set up the protocol for a dedicated port and actually *apply the protocol while we are modifying it* to add more details at development time.

Our device message on page 44 is a good start. The semantics dictate data coming from a device. Any developer can read the device message as "the data created on the device and delivered in a specified ORM structure". This information is sufficient to implement the protocol for Traccar to configure a port for the device protocol message – even before relating positions to it!

Keep in mind that we have put ourselves in the comfortable situation of working with the existing Traccar model defined in the database. Data modeling *is* programming and provides the base for our software development. Setting up a model from scratch requires quite some experience and a good notion of design impacts. On the other hand initial modeling can hardly foresee all system requirements and should be modifiable in the process.

For rapid development the architect hands over the device message (i.e. the proto file) to the device communication developer to set up the protocol. In the development cycle client, server and protocol are always compiled together by the build tool to propagate protocol changes. Once the protocol is in place the application developer can setup *his* environment to transfer the first messages. Then the latter can start modifying the protocol without any configuration changes.

As demonstrated with the battery level you should always thoroughly go through the required use cases when implementing new sensor data. The more complex use cases are those that should raise *NO-events*. In case of ambient assisted living the system should be able to raise an event, if the tracker *does not* send any positions for a specified time. This can be covered by switching on the trackers heartbeat messages. In SDC scenarios heartbeat messages are not applied and if a car doesn't send anything - who cares.

Now we will get back to the Traccar model to improve the device message. Every tracking system is based on the concept of client events where every event comes with a GPS(lat,lon,time) coordinate. The Traccar model implies a useful distinction of positions and events. Instead of using interval events (for time or distance) the positions are stored without event information. This makes sense for 'live tracking' which can be driven by a complex Client software logic. This way the tracking system is simply tracking the device without raising any events.

Now you, the network modeler, are facing the prize question: How do I model the device message to handle the relation device - position/s - event/s? The solution is simple:

```
message Device {
   int32  deviceid = 1;
   repeated Position position = 2;
}            --------

message Position {
      :
   double latitude  = 6;
   double longitude = 7;
      :
   repeated Event event = 9;
}            -----

message Event {
      :
}
```

Keep in mind that the event message should be filled on the client side while the Traccar GTS can generate server events! If you want to use predefined event keys, like the ones listed for Traccar on page 35, then you should *not* use strings to transfer over the network. With protobuffers you can also create enumerations which are much more effective:

```
message Event {
   EventType event = 1 [default = KEY_EVENT];
      :
}

enum EventType {
   KEY_EVENT    = 0;
   KEY_ALARM    = 1;
   KEY_BATTERY  = 2;
   KEY_IGNITION = 3;
   KEY_MOTION   = 4;
      :
}
```

The actual message design will be continued with the implementation of the

JEETS protocols project. Now we have defined a `1:m:n` relation without using position nor event `ids`! Their creation takes place in the database to recreate the relation when requested from the database with an SQL statement. We will see how the Java Persistence Architecture (JPA) can create a new ORM with identical content from system entities defined in a Persistence Unit (PU).

5.6 EXTENDING ORMS

After some decades software development has reached the remarkable state of *continuous testing, integration and deployment* which implies a constantly changing software (on the device) to avoid bothering users with update notifications. With the device message we are in good shape to extend the message with additional related messages. This section will describe a theoretical extension, which is currently not implemented, but should give you a good idea of the design process for whatever your system needs.

Most, if not every tracking system works with the concept of geozones to raise `entry` and `exit` events for a specified area. Of course it is much more user friendly to interactively draw geozones on a browser map frontend than creating abstract `(lat,lon)` tuples. On the other hand it may be necessary to transfer these geozones to a tracker. Similar to the battery level evaluations this allows the tracker to trigger these events.

In Traccar geozones are defined in a single string of Well Known Text[2]:

```
"POLYGON((49.03148673558755 12.100328001555415,
         49.03151487277535 12.104791197356196,
               :
         49.03148673558755 12.100328001555415))"
```

To handle geozones with Java you can define it in a proto message:

```
message Geozone {
   string name = 1;
   repeated Coordinate coordinate = 2;
}
message Coordinate {
   double latitude  = 1;
   double longitude = 2;
}
```

and you now be should able to figure out how to add it to the device message or to create a geozone message!

Note that we are using simple two dimensional coordinates instead of using the position message for tracking. Timestamps wouldn't make sense. Another important thing to consider is the *protocol direction*: Do you want to send geozones to the device or receive the devices geozones?

---

[2]en.wikipedia.org/wiki/Well-known text

5.6.1 Temporary ORM Extensions

To wrap up the protocol design we will describe how to add a network model extension for a dedicated project – without modifying the system's data model.

As mentioned earlier client software can constantly analyze the output of the GPS unit to improve the results to be transfered. As a GPS unit usually provides NMEA sentences as output, the client software can apply these sentences for the analysis. Now the client developer can not simply rely on the delivered positions without validating the results on the server side. Therefore he decides to add the NMEA sentences to the device message for his personal development environment without bugging the colleagues programming the main application based on the system's data model.

The geozone example provided rough guidance on how to add related messages to the device message. The client developer can add a NMEA message to the device message and implement a logic to evaluate the NMEA sentences and develop the client software on the server side which is much more convenient than doing an Over The Air (OTA) update over and over again. Once the software is in good shape the developer can remove the NMEA message again without any side effects on the system.

The network model and system data model should be well aligned at all times as they are not identical and can be modified for their respective environment, i.e. network model for networking only!

5.6.2 Harmful ORM Extensions

Creating protobuffer messages is simple and even simpler if you already have a relational data model to work with. Similar to creating a persistence unit (in the next chapter) one could be tempted to create a protocol for the complete data model at once. This is not useful and can even be harmful.

Recall that a tracker has to be registered before the tracking system will record its track. If you decide to widen the device message with every field available in the main model a junior developer might pass these messages to the system and implement an automatic device registration initially for his own purposes. The point is that you loose control of the systems integrity, if you don't perceive the protocol as the key to the system. Otherwise you would loose the servers power to administer the system.

Like the Traccar model you can usually subdivide sets of tables for different situations:

Administration - for people, devices, booking etc.

Incoming Data - from client software

Outgoing Data - for client software

Map Data - for calculations etc.

and for every category you should consider creating a message for a certain use case – or avoid sharing any system information in the protocol. For self driving car systems we can expect the car industry to come up with different payment models and then the server should only serve, if the customer has paid! With the JEETS components you will be able to hook up your CRM and debit the customer as needed.

We have now covered Traccar's relational data model and demonstrated how to create a relational network model, i.e. protocol. In the next part of the book we will create JEETS components, i.e. Maven Projects, to provide the Relational Data Model and Network Model for the system.

II

JeeTS Data Formats

So far we have gone through the concepts of message exchange and device communication with well defined protocols in conjunction with the system's relational data model. These two structures are fundamental to GPS tracking in general. Therefore we will catch up with the theoretical introduction by implementing these two core components.

In the Traccar source we can identify the `org.traccar.model` package as the relational data model and the `org.traccar.protocol` package as the decoders for various protocol formats and types. While Traccar is a monolithic Standard Java application JEE development is based on creating components which are combined to a software by a build tool.

We will create JEETS components for each programming domain: One component to provide the relational data model and another to handle associated protocol messages. Each JEETS component should include its dependent JEE component, i.e. Netty framework and protobuffer library, to implement test code on the component without any external dependencies. Since these components can be used for any application based on this model it makes sense to avoid as much business code as possible.

Initially JEE development was related to the application server as a single execution environment. The requirements of social networks, the self-driving car with Big Data etc. have changed this approach. While the JEE specification is taking care of the compatibility and coordination of its modular components it is not a software of its own. The idea of extending Java Standard APIs with higher lever functionality has turned out to point in the correct direction. In the context of JEE build tools the data- and protocol model can be perceived as drivers which should be applied by every developer to guarantee (data-) compatibility. Later we will see how to apply these modules for any (Java-) context you may choose.

CHAPTER 6

JⱻⱻTS Persistence Unit/s

CONTENTS

6.1   INTRODUCTION

The relational data model is the fundamental component for most systems and JᴇᴇTS components can only interact if they share the same data formats: Entities should be considered as the smallest and *related* pieces of information for the system. The power of Jᴇᴇ development lies in the clear separation of components for different programming domains. These components can be downloaded from third parties like the Netty Framework and the Protobuffer Library. And they can be combined with your business code to create a new component including the third party dependencies.

This chapter will deal with the `org.traccar.model` package representing the installed database. If you are familiar with the Java Persistence Architecture (JPA) then you already know that this package is defining the *Traccar Persistence Unit* (PU).

We have introduced the idea of Jᴇᴇ components in Section 2.3 and JᴇᴇTS should provide a number of useful tracking components. While Traccar is a well aligned Java application and only requires a JVM we will create a separate persistence unit for *any* application based on the Traccar model. A `jeets-pu-traccar` module can be used to replace the `org.traccar.model` package outside of the Traccar application – like in an application server or some (OSGi-) container. The PU provides all information to create a database, even in memory, instantly to be used by the environment, which might only exist for some dedicated processing steps.

The readers not familiar with JPA can look at a `pu.jar` similar to a database driver like `postgresql-9.4-1200-jdbc41.jar`. However the PU is including the *semantics*, the relational data model for our system. With this component we can develop JEETS components based on the Traccar model in conjunction with a database management system.

For creating a network model we have used the protocol buffer technology to setup a (complex) protocol message from the application's entities and buffer it until the software and network are ready to stream it. The JPA specification is based on a similar intention.

By providing entities for every table (class) and row (object) the developer can freely create an entity with the `new` keyword and assign the values to each member variable. In JPA the 'buffer' is provided by an `EntityManager` which *can be* involved to synchronize entities with a persistence layer. Since we relate a database engine to an external harddisc the database access can significantly slow down the performance. We will highlight some pitfalls and usage patterns in the Implementation phase of JEETS components.

6.2   JAVA PERSISTENCE ARCHITECTURE

The Traccar GTS does not apply a JPA implementation (`javax.persistence` packages). It applies the Standard Java `java.sql` package to compose SQL statements, HikariCP [2] to manage the connection pool and liquibase [3] as a change management (and migration) tool for SQL databases.

The Java Persistence Architecture (JPA) provides a higher level API than the classical JDBC approach as it can implicitly manage database synchronization with Object/Relational (ORM/ERM) Mapping. By providing a Persistence Unit (PU), which is basically a Java package with each class representing an entity of the data model, you can focus on your Java code and leave the database synchronization to an entity manager.

JPA is only a specification and in order to apply it it needs to be implemented. Similar to choosing Netty as TCP framework, there are various powerful JPA implementation solutions freely available. The Glassfish community usually uses EclipseLink [4] while the JEETS follows the JBoss Community and we will use Hibernate [20] to create two persistence units for the same data model.

Although we have not introduced any JEE environment yet it makes sense to create a JPA specified PU while playing with the model generator. The project `jeets-pu-traccar-jpa` is versioned after the Traccar release. Thereafter the architect is prepared to point developers to the data model. If the company guidelines strictly insist on JEE specifications the JPA module should be used.

In our analysis of relational modeling in Section 4.3 we have found Traccar's `Position` entity modeled with inheritance:

```
public class Position extends Message extends Extensible
```

The good news is that you can re/model the data model *inside* your Java code to introduce common objects and entities to inherit from. The important standardization is the access layer to the database and our new `Position` represents the same entity to be synchronized with the database (-table).

6.3 HIBERNATE ORM

In our terminology Hibernate is our next JEE specified component we will apply to create a JEETS persistence unit for the Traccar data model:

Hibernate Object/Relational Mapping [21]

> "Hibernate ORM enables developers to more easily write applications whose data outlives the application process. As an Object/Relational Mapping (ORM) framework, Hibernate is concerned with data persistence as it applies to relational databases (via JDBC)."

Relational data modeling is an architects core task throughout the complete project development and we are applying the Traccar GTS model to roll out useful JEETS components. By installing Traccar you (should) have created the PostgeSQL database which makes it easy to create a persistence unit with Java code. Hibernate supplies a set of tools to achieve this easily. This chapter will demonstrate how to create persistence units from a database.

Hibernate Tools [22]

> "Working with Hibernate is very easy and developers enjoy using the APIs and the query language. Even creating mapping metadata is not an overly complex task once you've mastered the basics. Hibernate Tools makes working with Hibernate or JPA even more pleasant.
>
> Hibernate Tools is a toolset for Hibernate implemented as an integrated suite of Eclipse plugins, together with a unified Ant task for integration into the build cycle. "

For JEETS development we will use the Eclipse IDE with Hibernate Tools:

Reverse Engineering [22]

> "The most powerful feature of Hibernate Tools is a database reverse engineering tool that can generate domain model classes and Hibernate mapping files, annotated EJB3 entity beans, HTML documentation or even an entire JBoss Seam application in seconds!"

We will not go through the procedure here as there are enough sources on the Internet. Here is a rough outline:

install Hibernate Tools in Eclipse

open Hibernate Perspective

add New Hibernate Configuration

add JDBC Connection to the Traccar database

add New Hibernate Code Generation Configuration

reverse engineer from JDBC Connection

select Export Folder to export data model to

When generating a persistence unit for the first time you should take your time to generate different models with varying options and compare the results. Basically you can choose from three models

Native Hibernate APIs and `*.hbm.xml` mapping
entities without annotations, `*.hbm.xml` mapping files
for details of table (cols) and DB (relations)

Native Hibernate APIs and annotation mappings
`javax.persistence` annotations are used to provide the metadata,
rather than a separate mapping file.
Hibernate Config files define classes instead of resource/s

Java Persistence API (JPA) via `persistence.xml`
to bootstrap a JPA `EntityManagerFactory`
Uses annotations to provide mapping information

In general Hibernate works with a proprietary implementation with pure Java code in conjunction with Hibernate mapping files `*.hbm.xml` as resources to define fields and relations. You can find the results in the project `jeets-pu-traccar-hibernate`[1]. The JEETS is modeled with JPA entities according to the JEE specification.

By choosing Hibernate as we are defining something like a pivot for persistence. Hibernate can be used for JEE environments with its own dialect or according to JPA specification. Beyond that the Hibernate product is keeping up with new development strategies like the

Hibernate Object/Grid Mapper (OGM) [23]

> "Hibernate OGM provides Java Persistence (JPA) support for NoSQL solutions. It reuses Hibernate ORMs engine but persists entities into a NoSQL datastore instead of a relational database."

6.4 JEE PERSISTENCE UNIT

The other PU resides in the `jeets-pu-traccar-jpa` directory and is a clean JEE specified JPA component. This implementation should fit some of your existing JEE applications and is not related to Hibernate any more.

JPA Provider [21]

> "In addition to its own 'native' API, Hibernate is also an implementation of the Java Persistence API (JPA) specification. As such, it can be easily used in any environment supporting JPA including Java SE applications, Java EE application servers, Enterprise OSGi containers, etc."

After running Hibernates reverse engineering tool you will note that the entity naming does not work well to assign the tables plural to singular for entities:

```
@Entity
@Table(name = "positions", schema = "public")
public class Positionses implements java.io.Serializable {
    ===
```

Therefore the PU was manually modified. Another crucial pitfall is the misalignment of getters and setters with their variable. You should be careful when you manually change a `position` to a `coordinate` for example.

```
public void setPositions(Set<Position> positions) {
    this.coordinates = positions; }
```

---

[1] Please note that this `jeets-pu-traccar-hibernate` sample project will not be maintained with new Traccar releases!

This *can* work in certain cases and certain implementations. Anyway it does not fulfill the *Bean Contract* and can raise problems when the JPA provider uses Java Reflection to align member variables with getters and setters. And we all know Murphy's Law...

6.5   FINE TUNING ORM AGAINST ERM

In addition the JEE persistence unit can be fine tuned to serve the application and reduce programming. On page 48 we have designed the `Device` message to fit incoming data to Traccar and to define a single protocol for related messages. When working with entities (with or without the database) we would like to introduce the same convenience.

Let's say one module creates different entities with the keyword `new`:

```
Device device = new Device  (1, "myTracker", "395389");
Position pos1 = new Position(1, device, .., 49.12d, 12.56d, ..);
Position pos2 = new Position(2, device, .., 49.34d, 12.78d, ..);
Event  event1 = new Event   (1, device, "deviceMoving", ..);
Event  event2 = new Event   (2, device, "deviceStopped", ..);
```

then the `Position` and `Event` entities are related to the `device` by using the object for the relation. Now these five objects can be handed to the entity manager for persistence in the database:

```
entityManager.persist(device);
entityManager.persist(pos1);
entityManager.persist(pos2);
entityManager.persist(event1);
entityManager.persist(event2);
```

This approach implies some problems. After creating the five entities the programmer has to make sure that they are persisted in the correct order of dependencies! If you wouldn't persist the device first and then store its positions and events this would raise a violation in the database as they wouldn't have a parent. It is also inelegant to invoke a `.persist()` method on every single entity. By doing this you create additional work for the entity manager. Let's see what we can do to move this persistence ordering and id management from the business code into the PU.

6.5.1   Hiding IDs

In the 'Data Modeling' chapter on page 36 it was stated that a programmer should not need to know nor touch any IDs. The above code with explicit IDs works for an empty database and the programmer can choose any identifier value to relate the entities. Naturally this approach is not fit for Enterprise development. Although the model generator provides entity constructors with explicit IDs

```
public Device(int id, String name, String uniqueid) { .. }
```

these should not be applied by developers. Yet they might be required for the Persistence Provider itself!

To get a grip on related entities it is very helpful, if not obligatory, to create test cases for each use case to be implemented in the system. It is also very helpful to provide data samples in the actual Java code to be used in `main(String args[])` methods of other projects. Have a look at the `org.jeets.model.traccar.util.Samples` class where we can `createDeviceWithPositionWithTwoEvents`. First we create to independent entities

```
Device device = createDeviceEntity();
Position position = createPositionEntity();
```

and then we relate them to each other

```
position.setDevice(device);
device.setPosition(position);
```

6.5.2 Creating IDs

On the other hand we know that database relations are based on IDs. If you open the database table 'positions' you can manually add a position *without* providing the ID and the store it. Then you will see a new ID and if you look into the relational model created by the Traccar installation in Postgres you will find a sequence generator

```
CREATE SEQUENCE public.positions_id_seq
    :
id integer NOT NULL DEFAULT nextval('positions_id_seq'::regclass)
```

while the entity simply adds the annotation `@Id` to direct the ID generation.

Sometimes the database model is provided by a database creation script and already includes sequence generators. Then the designer of the PU should align their names with the Java code with the annotations

```
@Id
@GeneratedValue(strategy = GenerationType.SEQUENCE,
                generator = "positions_id_seq")
@SequenceGenerator( name = "positions_id_seq",
            sequenceName = "positions_id_seq", allocationSize=1)
private Long id;
```

The JEETS PUs should be able to create the database in a production environment. In the testing section we will see how to create a complete database from the PU[2].

---

[2]with `hibernate.hbm2ddl.auto = create`

Similar to creating one single `Device` message for the network model to transfer related positions and events we want to introduce the same elegance to our database model. The software can create a `Device` entity with related children and pass this ORM through various methods which can read and modify every member on the way.

The Java persistence unit code does not require a database and the developer should know when to involve the `EntityManager`. Since we didn't like the persisting of every single entity in the first example we have tuned the PU to store a parent entity with all of its children. By adding a `CascadeType` to the entities relation in the code

```
@OneToMany(fetch = FetchType.LAZY, mappedBy = "device",
          cascade = CascadeType.PERSIST)
public Set<Position> getPositions() { .. }
```

we can instruct the device entity to store its positions, events etc. Now we can finally persist the device in a single line:

```
EntityManager entityManager =
                entityManagerFactory.createEntityManager();
entityManager.getTransaction().begin();
entityManager.persist(device);  // .. with all related objects!
entityManager.getTransaction().commit();
entityManager.close();
```

The first two lines (and the instantiation of the EM factory) can be anywhere in the programming logic and the last two lines should be used for a clean termination of the entity manager[3].

6.5.3 Object Relational Mapping

For this book we decided to perceive the Traccar software as an existing system, which is the normal situation in collaborative software development where you hardly get a chance to start a system from scratch. Therefore the ER model may already be in use on many productive servers, i.e. your Traccar installation, and does not allow any changes.

We have used Hibernate's database reverse engineering tool to generate entities for Object/Relational Mapping and now want to have a look at some details. First we'll have a look at some selected Traccar's database tables and relations

```
table     |    id relations
----------+----+----------+-----------
events    | id  deviceid   positionid
positions |     deviceid      id
devices   |        id      positionid
```

---

[3]Note that the explicit transaction control is not required in an EJB context.

We can derive

1. `events` are related to `positions` `n:1` via `positionid`

2. `events` are related to `devices` `n:1` via `deviceid`

3. `positions` are not explicitly related to `events`

4. `positions` are related to `devices` `n:1` via `deviceid`

5. `devices` are related to `positions` `1:1` via `positionid`

The additional uncommon relation in the last line is a special case of the GTS. The device's entity can store the last known GPS coordinate which is used to display a devices location on a map. The single `positionid` *does not* represent the relation of devices and their positions (`1:n`)! In our network model the device message does not have a direct relation of events to devices. We have devices with positions with events which is straight forward and easy to understand. In other words every event has to have a position coming from a device. From this short analysis we can conclude that an existing ERM is not necessarily normalized and needs some special attention when it comes to transformations etc.

Next we'll have a look at the Object/relational Mapping created by Hibernate and figure out some typical use cases of the system.

```
Entity     | Relations
-----------+-----------------------------------------
Device     | Set<Event>    Set<Position>    positionid
Position   |                   Device
Event      |    Device                      positionid
```

What can we see in this table? We can see that the device entity *does have* direct relations to events and positions which can be used in the code

```
position.setDevice(device);
device.setPositions(Set<Position>);
device.setEvents    (Set<Event>);
entityManager.persist(device);
```

This will be sufficient for a protocol to PU transformation and for a cascaded persist of the device in one method call.

The special case of the `device.positionid` described for the table relations was modeled with

```
public class Device {
   private Integer positionid;
   public  Integer getPositionid() { .. }
   public   void   setPositionid(Integer positionid) { .. }
```

Hibernate has generated getter and setter for an explicit ID. This is problematic when working with entities without database and should be modified to[4]

```
private Position position;
public  Position getPosition() { .. }
public    void   setPositionid(Position position) { .. }
```

in order to relate the device to the last known position. The event entity has the same issue and should also be modified to set its GPS position.

In the little overview table we can see that the position entity has no relations to events. The ERM only relates events to positions with events.positionid, not vice versa! In our programming environment it would be convenient to create events as indicated in Figures 2.1 and 5.1.

```
Event moveEvent = createEventEntity();
moveEvent.setType("deviceMoving");
Event stopEvent = createEventEntity();
stopEvent.setType("deviceStopped");
```

and then we relate them to one position

```
position.setEvents(Set<Event>);
```

instead of

```
moveEvent.setPosition(position);
stopEvent.setPosition(position);
```

Although the positions table has no ID to the events table we *could* add a member to the position entity

```
private Set<Event> events;
@OneToMany(...)
public Set<Event> getEvents() {
   return this.events;
}
public void setEvents(Set<Event> events) {
   this.events = events;
   for (Event event : events)
      if (event.getPosition() == null)
         event.setPosition(this);
}
```

By adding some logic to the event setter we could set the event's position implicitly and without IDs! Anyway this kind of change can raise logical problems that might initially not appear in some simple test cases and once you release them to your team you might introduce these problems to many projects relying on a well structured PU.

---

[4]Should be applied with an associated use case.

6.5.4   persistence.xml

The line

```
EntityManager entityManager =
               entityManagerFactory.createEntityManager();
```

defines the JPA bootstrap process with its own configuration file named
`persistence.xml`. In Java SE the persistence provider is required to locate all
JPA configuration files by classpath lookup of the `META-INF/persistence.xml`
resource name. Inside the code the PU is referred to by the

```
<persistence-unit name="jeets-pu-traccar">
```

with its name string

```
entityManagerFactory =
   Persistence.createEntityManagerFactory("jeets-pu-traccar");
```

Since the `persistence.xml` file is distributed with the PU project you
should consider if every property is really fit for global usage. For one you
can create a separate `persistence.xml` for testing. On the other hand you can
create a minimum version with the database driver

```
<property name="javax.persistence.jdbc.driver"
        value="org.postgresql.Driver" />
<property name="javax.persistence.jdbc.url"
        value="jdbc:postgresql://localhost:5432/traccar" />
<property name="javax.persistence.jdbc.user" value="postgres" />
<property name="javax.persistence.jdbc.password" value="postgres" />
```

and then add the properties via programmatic configuration depending on the
JPA provider. For EclipseLink it would look like this:

```
java.util.Map<String, String> entityManagerProps
= new HashMap<String, String>();
entityManagerProps.put(
   PersistenceUnitProperties.TRANSACTION_TYPE, "RESOURCE_LOCAL");
   :
EntityManagerFactory emfactory = Persistence
        .createEntityManagerFactory(PU_NAME, entityManagerProps);
EntityManager emgr = emfactory.createEntityManager();
```

For Hibernate

```
Configuration cfg = new Configuration()
   .setProperty("hibernate.dialect",
              "org.hibernate.dialect.MySQLInnoDBDialect")
   .setProperty("hibernate.order_updates", "true");
```

In the `PersistenceTest` you can find a way to override the Postgres connection with an in memory database for testing.

We have generated the complete ERM to a complete ORM while only looking at three tables. Therefore we could create a persistence file with only three entities. Note that you can provide a dedicated persistence file in your higher level project to explicitly define the entities involved in your code. Whatever procedure you choose the most important part is to include the classes in the persistence file and `<exclude-unlisted-classes>` as you can see in the provided `persistence.xml` files.

This section has introduced a JEE JPA specified persistence unit and demonstrated some tuning you can do inside it. Another thing you can do or find in the entity classes is the definition of named queries that developers can use to create data Access Objects (DAOs) without dealing with low level SQL. Now let's see how we can actually 'run' the PU to create and use a database in the test environment.

6.6   PU TEST ENVIRONMENT

The introduction of the terms JPA, PU, ORM and ERM was very lengthy and maybe the details are not of great interest to you. We have only analyzed three tables to indicate some problems you can run into if you simply trust the ERM to ORM automated reverse engineering.

The better way to experiment with entities and experience 'minor' problems is to implement use cases in the test environment[5]. The main message of the persistence analysis is that every use case concerning handling of entities in the code and synchronization with a database should be implemented as a test case. Whenever the ORM is modified you can run the tests to see the impact. Then you must increment the PU version and spread the news to your colleagues to have them increment the PU Version in their environment. On the other hand every problem with the PU occurring in another (higher level) project should be fixed in the PU! Although this procedure is simple it is often ignored in programming teams and is a main source of hidden costs due to complicated debugging analysis, etc.

The Maven project `jeets-pu-traccar-jpa` defines the first JEETS component and is fundamental for further development. As described earlier you change to this directory to invoke Maven goals to `clean`, `compile` and `test` the code at development time.

Generating the persistence unit/s from a JDBC connection is convenient – *initially*. As the Traccar database can vary with every release[6] the model has to be modified and you should define a strategy to generate a new model for every new version or to manually adopt the changes. We will define our update strategy in the next section.

---

[5] usually placed in the folder `src/test/java`
[6] the Traccar folder /schema includes the changelog xml files

The architect could now distribute the PU to different developers... can he rely on the full functionality? Not yet. In times of continuous integration every module should come with a set of tests to run after every modification. Testing with databases used to be a real problem and the JPA approach has provided powerful helper technologies.

Please open a command line and navigate to the directory

```
...repo.jeets/jeets-persistence/jeets-pu-traccar-jpa
```

Now you can invoke the Maven life cycles (one by one or all at once):

```
mvn clean
mvn compile
mvn test
```

If the project is in a clean state you should achieve a

```
[INFO] ----------------------------------------------
[INFO] BUILD SUCCESS
[INFO] ----------------------------------------------
```

What you see in the

```
----------------------------------------------
  T E S T S
----------------------------------------------
```

section is the complete creation of a database with the h2 in-memory database. First the database is cleaned up

```
Hibernate: drop table public.positions if exists
```

then it is being setup from scratch

```
Hibernate: create table public.positions (id integer ...
```

To persist the device entity with position and event children described in section 6.4 open the `PersistenceTest.java`[7] in your IDE and run it from there. A test suite is much more than a black box test. Developers can look into the test code to transfer patterns to their own code. On the other hand each developer is responsible to implement a test case for every entity.

Note that, due to constant Traccar Upgrades, we are using the original `persistence.xml` and only modifying the database connection. In our case this is the safe approach to ensure that our software is using the same entities (and ignoring the irrelevant ones).

You can also check the console output of the `PersistenceTest.java` to find that inserting the device entity invokes the statements

---

[7] ../jeets-pu-traccar-jpa/src/test/java/org/jeets/model/traccar/jpa/

```
Hibernate: insert into public.devices (..) values (..)
Hibernate: insert into public.events  (..) values (..)
Hibernate: insert into public.events  (..) values (..)
Hibernate: insert into public.positions (..) values (..)
Hibernate: insert into public.positions (..) values (..)
```

to insert the dependent children one after the other.

Developers are responsible for creating a test case for every new implementation. The test case should focus on testing entities and their relations against a database. Technically speaking there should be test cases with database Creation (DDL – Data Definition Language) and Data Manipulation (DML – Data Manipulation Language).

One of the rapid development techniques is to start the implementation in a test case. This is very convenient and makes sense. Anyway the developer should *never copy* from a test environment and paste the code in his environment. He should rather *move* the code to the system implementation and then reduce and redefine the test case. In a higher level component the test case might simply invoke the new method to call the recent test code. And for data modeling the developer should implement the data access in the persistence unit tests.

As another development practice we have added a test class for `PostgresTesting`. Due to the naming unequal to `PostgresTest` the test is not executed in the context of the continuous repository build as it requires an existing Postgres DBMS. Anyway a developer can create a Postgres test database and experiment with database creation by invoking the test from his IDE.

In a later stage a second repository run can be created to simulate a complete process with a real postgres database. This is usually added later when more interdependent modules exist. Then the complete process is kicked off with a Maven switch like

```
...\repo.jeets> mvn clean install '-Prequires-postgres'
```

to invoke the Maven profile with this name.

```
<profile>
<!-- All the quickstarts that require Postgres to be running -->
  <id>requires-postgres</id>
```

6.7   ERM AND PU UPDATES

At this point we have generated the complete persistence unit with all Traccar tables. Then we have modified the three related entities `Device`, `Position` and `Event` and added some JPA features that support our code (like the `CascadeType`).

We have put ourselves in the situation to develop new components for a running production system. 'Production' usually refers to an ongoing business.

Consequently we have to keep up with Traccar's development and apply the ERM updates with every new release. A closer look at the database schema history reveals that tables can be dropped from the model and new tables and attributes are added constantly.

How can we deal with this constant development? Software is alive and developers should be aware of this fact. How did the creators of the JPA specification incorporate this aspect?

For one thing the direct access to the *database* bypasses the logic of the Traccar *software*. Consequently the JEETS modules can insert any device with positions and events as long as they adhere to the table's constraints. We will create a JEETS loader to do exactly this and implicitly take over the responsibility for the values to work with the Traccar software. For example we can add a validity check to only insert a devices data – if we can previously look up the device in the table. In other words the device has to be registered via Traccar before the JEETS loader can add positions and events.

Another important aspect of designing a dedicated persistence unit *for a piece of software* is that the PU *does not* have to include the complete ERM. As we have only modified three tables yet we can create the PU with only these three tables! And we can create additional PUs with different tables.

If we want to add (geofence) logic as described in Section 5.6 we can add the relevant tables and entities in the development process[8]. This way the maintenance can be handled much better. Generally it is easier to add than modifying existing tables and attributes. Even if the existing tables are extended the entities only have to modified, if they handle the changes.

Practically the PU is regenerated after every new Traccar update and moved into the jpa folder. We have renamed the three entities from plural to singular. Therefore they will not be replaced, the new software will work with the manually modified entities. The other entities are compiled anyway and can be modified as they are implemented in the JEETS software.

For large developer teams it can make sense to create different PUs for different purposes:

```
jeets-pu-traccar-devices-jpa
jeets-pu-traccar-admin-jpa
 etc.
```

With the JEETS approach to use Traccar's frontend for administration of devices and people we can skip the majority of tables and easily update every Traccar release. In order to ensure that only the manually created entities are applied in a software module is to explicitly place the entities in the `persistence.xml` and add the line:

```
<exclude-unlisted-classes>true</exclude-unlisted-classes>
```

For a quick start of the new system we are ready to go and have looked at

---

[8]which will be described in Section 17.4.2

some fine tuning options of a persistence unit at development time. Database testing has always been problematic in terms of performance. With modern build tools it is easy to use an embedded database for testing. On the other hand it can be cumbersome to insert enough data for performance testing. Of course a test environment can also access a DBMS which was demonstrated in the test class `PostgresTesting`. By renaming it to `PostgresTest` it will be executed for every `mvn install`. This is fine at development time, but should not be released to the official repository. The problem lies in the unknown status of a database in a continuous release process. While the developer can search the (production) system for relevant datasets the integrator might set up a different database for testing and the datasets won't be found.

If you want to focus more on PU tuning you can continue by reading Chapter 17.4 and then come back here. There we will have another look at the relationship of Java code to performance tuning for a production system. We will inspect how we can optimize performance with SQL and JPQL according to the Java implementation for ORM navigation.

JeeTS Protocols and Decoders

CONTENTS

In this chapter we will introduce the protocols module `jeets-protocols` for all JEETS modules. In Traccar all protocols lie in the `org.traccar.protocol` package and you can study each parser to adapt it to you needs. For JEETS development we will use protobuffer protocol files to design protocol messages associated with the relational data model in the PU.

7.1   COMPILE TRACCAR.PROTO

In the chapter 'Object Relations' we created a basic Traccar message (see page 48) to submit GPS coordinates and client events from a device:

```
message Device {
    string uniqueid = 1;
    repeated Position position = 2;
}
message Position {
    double latitude  = 6;
    double longitude = 7;
        :
    repeated Event event = 9;
}
message Event { .. }
```

In order to compile the protocol file you need to install the protobuffer compiler `protoc` in your environment. You might wonder about the extra effort to compile the proto file before you can start programming. There *are* ways to include this compilation in the Maven build process. If you think about it in detail: any change of the proto file requires some kind of coding. You could add a message or field and successfully compile the proto file without errors. Yet this can not implement the new structures in your application by itself. Therefore each developer should submit every proto change with the code changes to the versioning system.

7.2 TRACCAR.JAVA

If you have executed Google's person example [18] you can now compile the `traccar.proto` file and open the generated `Traccar.java` file. (You might want to backup the existing `Traccar.java` file first!) Don't get confused by finding a file with more than three thousand lines! The protobuffer library is a complete software to transfer Java objects over the network. The generated file is dedicated to the proto file, which is build on top of the library. We will not go into the details and you can study everything you need to know on the Protobuffer website. For our development we are only interested in looking at the member class `Position`

```
public static final class Position
    extends com.google.protobuf.GeneratedMessageV3
    implements PositionOrBuilder
{
    public double getLatitude() { return latitude_; }
        :        :
```

This `Position` class is the entity of the network model designed after Traccar's `org.traccar.model.Position` in the relational model. What we see is the position *received* from the network and therefore it only has getters and no setters. Transforming the network `Position` to `org.traccar.model.Position` is as simple as[1]:

```
Traccar.Position protoPosition = (Traccar.Position) message;
org.traccar.model.Position entityPosition = new Position();
    :
entityPosition.setLatitude(protoPosition.getLatitude());
    :
```

First you cast the received message to a proto position object, then you create a new Traccar position and assign every field which matches by design.

Of course you could also imply format changes, like `(d)dd.mm.mmmm` to decimal degrees. Then you should be aware of the performance for decoding

---

[1]see code in `jeets-tracker`

many coordinates on the server side instead of decoding a few coordinates with the client software. For SDC scenarios the client can take over many tasks to relieve the server.

7.3 PROJECT DEPENDENCIES

The network protocol will not and should not be part of the core GTS server and is only applied for device communication. Once the proto messages are extracted and transformed to Java objects there is no more need for protobuffers. Therefore the `jeets-protocols` project should include the Protobuffer Library and every developer can add the `jeets-protocols.jar` to his environment. Without knowing much about protobuffers the (not network) developer can directly access the network entities and *read* their fields with getters. Another good practice is to copy the proto sources into the archive where every developer can open them in his environment. Actually this is the only information the developer needs! You can control this behavior in the `pom.xml` file:

```
<build>
    <!-- Useful in development time. Remove for deployment? -->
    <sourceDirectory>src/main/java</sourceDirectory>
    <resources>
      <resource>
         <directory>protobuffers</directory>
      </resource>
    </resources>
</build>
```

In addition we shouldn't forget that the protocol was designed after the ORM represented by the `jeets-pu-traccar-jpa` project holding the entities. This means that the protocol project is not as isolated as the PU project and its structures should always be aligned to the persistence unit. Consequently we will import the PU into the protocol project

```
<dependency>
    <groupId>com.google.protobuf</groupId>
    <artifactId>protobuf-java</artifactId>
    <version>3.1.0</version>
</dependency>
<dependency>
    <groupId>org.jeets</groupId>
    <artifactId>jeets-pu-traccar-jpa</artifactId>
    <version>3.10.2</version>
</dependency>
```

When you run

```
jeets-protocols> mvn clean install
```

to create a `jeets-protocols-x.y.jar` you will note two tests:

```
------------------------------------------------------
T E S T S
------------------------------------------------------
Running org.jeets.protocol.TraccarTest
Running org.jeets.protocol.TransformerTest
nter
```

The second test class is there to ensure the compatibility of protocol and persistence data and we will now have a look at some implications in the next section.

7.4   DATA TRANSFORMATIONS

The protocol messages were designed after the persistence unit and naturally both belong together. Obviously the protocol is basically for device communication, the communication of a tracker and a small software, like a Netty decoder, to pick up the messages on the server side. We have gone through this process inside the Traccar GTS. We have also pointed out that every incoming message is transformed into Traccar's position entity.

Since the protocol project includes the persistence unit it makes a lot of sense to add their transformations to this package. This way each developer can use the project to have access to protocol messages and system entities. He can transform them without going through the details in every new component. Another advantage is that you keep the data transformation in a central place and simply import it any number of times.

The project jar should be used in all networking applications for the Traccar model, or more precisely for the counter parts of client *and* server. During development the Traccar protocol file can be enriched with more messages for additional use cases and the file represents a complete message catalog. On the other hand the persistence project can be used inside the server software, like an application server, without carrying the network stuff with it.

Please open the `org.jeets.protocol.util.Transformer` to find tranformation methods for `protoToEntity*` and `entityToProto*`. All of the methods are tested with `..util.Samples` in the `..util.TransformerTest` class. If you plan to create your own protocol you should setup a similar structure.

Some remarks on the implementation:

1. The transformation in the production environment is designed for one direction. By creating transformation tests you will run into these details and will implicitly prepare the use cases for higher level developers.

2. As we have analyzed in Section 6.5.3 the position entity has limited relations to events and device. This issue was not improved since we declared the device message as our protocol type. We will not transmit

single position messages, but rather device ORMs with positions and events included.

3. The device entity and device message are not congruent. Each has its explicit fields for their domain. For example the servertime can not be set by the client.

4. The `entityToProto*` methods do not return a protobuffer message. They provide the *modifiable* version: a protobuffer builder. This builder can be modified after the transformation. For example a tracker software will set the `deviceTime` only just before sending the message. If the transmission fails he can set the timestamp again and `.build()` the message for networking.

5. In addition to device, position and event the protocol message was modeled with event types. You can use the transformation process to convert Strings to Java enum types or transform in a more complex way. As an example you can look at the enums `EventType` and `AlarmType` in the `traccar.proto` file.

   Another aspect of event types is their origin. All events from client to server should be transformed. On the other hand server events should not be transformed.

   Also be aware that our production system, the Traccar GTS, should keep on working and create events. And these events have impact on other functionalities like reporting, mapping, sending notifications etc.

   We will have a closer look at events in the tracker chapter.

7.5 TRACCAR PROTO DECODER

We have analyzed the Traccar device communication and have modeled the protobuffer communication with the Netty framework. Now we need to prove that we can actually communicate with a tracking system. Therefore we will now create a protobuffer decoder for Traccar before we can create a client, a tracker to send our protobuffer messages.

7.5.1 Traccar Modifications

We have gathered all pieces to prepare Traccar for protobuffer handling. In order to use your personal Traccar installation with a protobuffer decoder we will add the well known pair of classes to the protocol package

```
org.traccar.protocol.ProtobufferDeviceProtocol
org.traccar.protocol.ProtobufferDeviceDecoder
```

Since Traccar is constantly under development the two `*.java` files are placed in `jeets-protocols/src/main/resources` where they are *not* compiled as they depend on many Traccar classes and JEETS will use Netty 4 (`io.netty`) while Traccar is working fine with Netty 3 (`org.jboss.netty`).

Now you can add the decoder in four steps:

1. copy the two `ProtoBuffer*.java` files to the `org.traccar.protocol` package directory

2. add jeets-protocols to `pom.xml`[2]

```
<dependency>
   <groupId>org.jeets</groupId>
   <artifactId>jeets-protocols</artifactId>
   <version>${jeets-protocols.version}</version>
</dependency>
```

3. add protocol to Traccar's `default.xml` file

```
<!-- PROTOCOL CONFIG -->
<entry key='pb.device.port'>5200</entry>
```

The protocol name is hardcoded in the protocol file

```
public ProtobufferDeviceProtocol() {
   super("pb.device");
}
```

and has to match the entry of the configuration file.

4. register the `uniqueId` you are using in the client software

If your Traccar installation compiles successfully you are prepared to receive `Device` protobuffers with Traccar. First, let's review the code to see what happens.

7.5.2   Implementation

In the protocol file you will find the pipeline setup for protobuffer handling which includes the decoder file to receive the extracted protobuffer message. The initialization takes care of registering the protocol to Traccar's list of tracker servers:

```
public void initTrackerServers(List<TrackerServer> serverList) {
   serverList.add(new TrackerServer(new ServerBootstrap(),getName()){
```

---

[2]with `<enforceBytecodeVersion><maxJdkVersion>1.8</maxJdkVersion>`

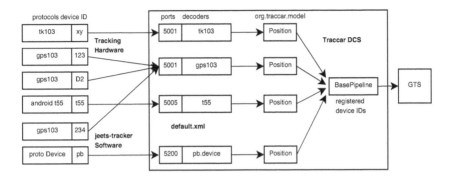

Figure 7.1 With the JeeTS-tracker software you can simulate any protocol encoder. The protobuffer device decoder creates system entities and works seamlessly with the Traccar DCS.

The decoder receives the messages in the method .decode where it can be directly casted to the Device message:

```
protected Object decode(channel, remoteAddress, Object msg) {
    Traccar.Device protoDevice = (Traccar.Device) msg;
```

and we can use the code we have already developed to

```
transformDeviceProtoToPositionEntities(protoDevice, deviceId)
```

Then we can create an Acknowledge message provided by the JEETS protocol jar and send it back to the client with

```
channel.write(ackBuilder.build())
```

And finally we return one org.traccar.model.Position or a List<Position> which will be processed down the Traccar pipeline to end up in the database.

Now we are ready to create client software to send these messages.
Let's create some client software to support development
and to observe a real track on Traccar's map frontend!

III

JeeTS Client Software

CHAPTER 8

The JeeTS Tracker

CONTENTS

8.1 TRACKING DEVICES

At this point we have done fundamental research but we haven't even trans-fered a message via TCP yet. We will now begin to roll out some components to work with, Software you can actually run to support client server develop-ment.

What exactly is a tracking device and what is its core functionality?

A tracker hardware is basically a combination of a GPS (processing) unit and a communication unit. The GPS *receives* the location via satellite and the communication unit *sends* tracking messages to a server. For developers the communication can be abstracted to TCP which was covered in earlier chapters. A tracker can use many transports to send messages, but in the end they arrive at an `IP:port` and are processed with an TCP Socket.

Since most trackers work with a GSM network we can speak of a GSM communication unit – while the software is actually based on TCP communi-cation which also includes any other network technology. Global positioning

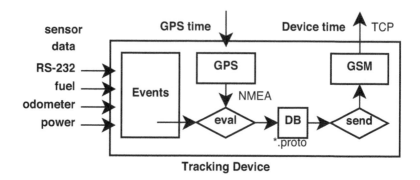

Tracking Device

Figure 8.1 A tracking hardware combines a GPS- and GSM unit on a Controller Board. The Controller evaluates sensor data to create event messages in the protocol format. NMEA data is constantly validated to provide the best position available. The messages can be sent whenever appropriate.

does *not* necessarily imply a GPS- and/or GSM unit. A tracker can be built explicitly for indoor tracking and communicate with different technologies than GNSS and GSM (like Bluetooth, WLAN and many more). In the end every tracker should be able to send Coordinates, preferably WGS84 latitude and longitude, which the Server can use to localize the device.

Many trackers need to operate *without* external energy source and are used to track people and assets. GPS and GSM are soldered on a single board with a controller logic responsible to make optimal use of the processors. The size of the tracker is defined by the circuit board's area plus the volume and weight of the accumulator or battery. A rechargeable battery with 1 Ah has the size of a matchbox.

GPS (or GNSS) receiver and GSM transmitter are standard electronic hardware modules which operate 'out of the box' and with the GSM you also gain a (personalized) SIM card and a phone number. The tracker vendor is adding a Controller to combine the modules operations for dedicated use cases. To face the main concern of energy management and consumption these trackers have sensors for motion (or acceleration to detect a falling device) to turn off the modules while the person is at rest. A 'smart' tracker is also able to detect indoor locations by evaluating the number and constellation of satellites, GSM signal strengths and triangulation etc. – and turn the GPS processor off for a while.

We are mostly interested in Vehicle trackers where Power management is not a real issue (while driving). The main difference is the connection to the Car's Data via CAN bus which requires a hardware installation to the On-Board Diagnostic (OBD / since 1996 OBD II) port – the Car's Computer! Of course the Car Companies developing their own SDC technology have access to much more than the official ODB II specification. Fleet Management systems can apply highly sophisticated devices to observe the vehicles 'health', maintenance intervals and much more.

8.1.1 Tracker Events

Better trackers use an internal database to store positions from the GPS unit whenever the tracker is disconnected. With this internally stored *Track* the controller logic can derive a number of events or status messages, which can be reported to the tracking system.

Here is a short excerpt of basic tracking events:

time / distance intervals

speed threshold (pedestrian or passenger)

angular change (course over ground)

power on/low/off

motion start/stop

alarm button pressed

Some more sophisticated trackers provide storage for geofences to trigger entry- and exit events *on the client side* !

8.1.2 Traccar Events

So far we have covered the functionality of the Traccar `protocol` and `model` packages which provide tracking data in the systems ERM format. Now we will look at the `events` package to understand how events are created in Traccar.

`Position` entities are created by protocol decoders inside the Netty inbound stream. Traccar is a general purpose GTS and provides more than a hundred message decoders for roughly a thousand different trackers and many different contexts. Consequently the `Position` provides a large number of event keys which could require different handling in the business logic.

Let's look at Traccar's `Gps103ProtocolDecoder` and see how events are handled.

```
parser = new Parser(PATTERN_OBD, sentence);
if (parser.matches()) {
           :
    position.set(Position.KEY_ODOMETER, parser.nextInt());
    position.set(Position.KEY_FUEL_CONSUMPTION, parser.next());
           :
```

The decoder stores each event as a key-value pair and the keys are provided by the system, i.e. `KEY_ODOMETER`. Some keys have subcategories like the `KEY_ALARM` which is subdivided by the method `decodeAlarm( alarm )`

```
private String decodeAlarm(String value) {
    if (value.startsWith("T:")) {
        return Position.ALARM_TEMPERATURE;
    } else if (value.startsWith("oil")) {
        return Position.ALARM_OIL_LEAK;
    }
    switch (value) {
        case "help me":
            return Position.ALARM_SOS;
            :
```

In order to create client events for Traccar the software should use these predefined values to simplify encoding and decoding. Now we can continue designing the Traccar protocol by filling in the event message we have already created

```
message Event {
    EventType event = 1;   // [default = KEY_EVENT]
    AlarmType alarm = 2;
}
enum EventType {
    KEY_NO_EVENT = 0;
    KEY_EVENT = 1;
    KEY_GPS = 2;
    //  set detail with AlarmType
    KEY_ALARM = 3;
        :
}
enum AlarmType {
    ALARM = 0;
    ALARM_GENERAL = 1;
    ALARM_SOS = 2;
        :
}
```

Note that the integers provided for each key (KEY_GPS = 2) only apply inside the protobuffer logic and are not related to the strings applied inside the Traccar model (KEY_GPS = "gps").

With this construct a *client* can create Traccar events with a protobuffer message:

```
private Traccar.Event createProtoEvent() {
    Traccar.Event.Builder eventBuilder = Traccar.Event.newBuilder();
    eventBuilder.setEvent(Traccar.EventType.KEY_ALARM);
    eventBuilder.setAlarm(Traccar.AlarmType.ALARM_SOS);
    return eventBuilder.build();
}
```

and we can directly decode the alarm type on the *Server* side

```
Traccar.Event event = protoPosition.getEvent(0);
Traccar.EventType eventType = event.getEvent();
if (eventType == EventType.KEY_ALARM) {
   if (event.getAlarm() == AlarmType.ALARM_SOS) {
      entityPosition.set(Position.KEY_ALARM, Position.ALARM_SOS);
   }
}
```

Again, this is only one way to model the protocol towards Traccar. Another way would be to move the detailed alarm types directly into the event types. You should decide whether your client software can only create one event per message... and so on.

Another important Traccar feature is the additional creation of system-, i.e. server events like the `MotionEventHandler` which is invoked inside Traccar's `BasePipelineFactory` to analyze the latest position

```
public class MotionEventHandler extends BaseEventHandler {
      :
   @Override // .. BaseEventHandler
   protected Collection<Event> analyzePosition(Position position) {
      :
   .. new Event(Event.TYPE_DEVICE_MOVING,
         position.getDeviceId(), position.getId()));
```

The `Event` is an entity of the Traccar model and can be persisted directly into the database table

```
CREATE TABLE public.events
(
   id integer ... ,
   deviceid integer,
   positionid integer,
      :
   CONSTRAINT fk_event_deviceid FOREIGN KEY (deviceid)
      REFERENCES public.devices (id) MATCH SIMPLE
      :
)
```

Looking at the event relation/s in the database definition we can find that the data model requires a registered device for every event. We don't find an explicit position-event relation? And the event does not require a position by the data model!

8.1.3   JEETS Events

Traccar's concept to analyze incoming positions directly in the Netty pipeline is the fastest and most effective way. Yet JEETS Components are created with respect to a self driving car context and need more freedom to create new events as they occur in the server backend. System events, like `AccidentAhead`, can only be created from a complex analysis of many cars, while Traccar handles each message in its own isolated channel pipeline. An SDC server should receive and extract plain event data and immediately return an `ACK` to release the pipeline, while the event continues to pass through the server logic.

Generally speaking we want to provide (`Position` etc.) entities to a system serving many different SDC scenarios. A main guideline is to process new messages inside the software (RAM) without taking the risk with slow or busy resources, like an external database connection.

By using the `Device` message we have created the possibility to send any number of positions and events in a single message. The client software can use some of its CPU power to optimize these small 'routes' for server processing. And the server's device communication should be able to propagate the incoming information to the system as fast as possible. Therefore we will create a JEETS 'loader' component in the next part of this book.

8.2   TCP SOFTWARE

In order to create a tracking software we'll state this simple use case:

"Whenever we want to track something, a person, pet, asset or vehicle we can equip this 'real world object' with a tracker to submit localized tracking events to a server."

The idea of the JEETS tracker is to wrap core tracking functionality in a single component, which can be used to 'equip' some thing. For example the tracker component could be installed on an Android client which has GPS and GSM hardware and provides a Java VM with Java libraries[1]. By simply reading the location information from the system they can be transformed, if necessary, and passed to the tracker software.

Again we can be happy to have Traccar and peek into its Android client implementation. Android provides the necessary accessors in Java and a tracking application can apply the service classes, like these

```
import android.location.GpsStatus;
import android.location.Location;
import android.location.LocationListener;
import android.location.LocationManager;
```

Remember that we are creating a tracker software and a *software* alone is not able to detect satellites or send something over the air. Location sources are accessed in different ways in IOS and many other systems. We can conclude that our tracker software simply needs setters for the GPS coordinates `lat`, `lon`, `alt` and `time`.

---

[1]developer.android.com/training/location/index.html

As we have introduced the `tk102` protocol in the message exchange chapter as a typical 'single string' protocol we will implement one of these most simple protocols to get started, before we will turn to the more sophisticated protobuffer communication.

We are going through the process of designing a system. The architect is basically putting all components 'in place' where developers pick them up for detailed implementations. You can use the component prototypes to implement your own (additional) logic. As an exercise you could implement an angular change event with an angle parameter.

8.2.1   String Message

The single string tracker should serve the proof of concept by creating a GPS message in a single character `String` and submitting it to a tracking system – in our case Traccar. This simple and rapid prototype is helpful for server development and can always serve as the fallback solution to the submission of a protobuffer message.

We have already introduced the plain TCP code on page 16 and you can find the code in the JEETS tracker project in

```
Tracker.transmitProtocolString(protocolMessage, host, port)
```

to find three lines of code to send a string to a host and port:

```
public void send(String positionMsg) {
    Socket serverSocket = new Socket(host, port);
    PrintWriter out =
        new PrintWriter(serverSocket.getOutputStream(), true) )
    out.println(positionMsg);
}
```

There are actually some more lines to wrap the connection in a 'try-with-resources' statement to catch connection problems. Yet the method has no return value and doesn't expect any return value from the server – send and forget. In order to test the `SingleStringTracker` with the Traccar server you can find many strings in the `../traccar3/server/test` directory[2] like this one for `gps103`:

```
'imei:123456789012345,help me,1201011201,,F,120100.000,A,
    6000.0000,N,13000.0000,E,0.00,;',
```

Modern Java EE development boils down to the composition of one software from many Components (with sub-, subsubcomponents etc.) with a built tool like Maven. The architect should provide the initial Maven projects and wire them together in the repository. The repository's project object model becomes *the* main software and developers add the business logic to individual components.

---

[2]or you can checkout the integration tests at
... tananaev/traccar/master/tools/test-integration.py

After creating a `jeets-protocols` project we will now create a `jeets-tracker` project 'on top' of it. In JEE terminology the tracker- depends on the protocol project. After a successful transmission of the `SingleString` message the architect can set up the Maven project and add a `jar-with-dependencies` compilation for a stand alone test.

8.2.2 Transmit String Message

To represent a simple external environment the `org.jeets.tracker.Main` class was added to actually invoke the tracker. You can use the `Main.main` method to send the message to Traccar. Remember to register the `deviceId` first, to supply the port on the client side and activate the protocol on this server port! Then go to the `jeets-tracker` project root to compile the tracker and immediately run a 'ping' to the GTS from the command line (replace `x.y` with the current version)

```
>mvn clean install
>java -jar target\jeets-tracker-x.y-jar-with-dependencies.jar
        localhost 5001
        "imei:359587010124999,help me,1201011201,,
         F,120100.000,A,6000.0000,N,13000.0000,E,0.00,;"
```

to send your message and check the server database and frontend. If you don't have your Traccar system up and running Traccar also provides a demo platform at `demo.traccar.org` where you can send your string once you're sure it works. For multiple tests you can also use a batch file and start with the `sendMessage.bat` file supplied in the main directory. Now the tracker software can be used stand alone to send `SingleString` messages to different ports. The project can be released to the repository where it will be picked up by a protobuffer developer.

8.2.3 Protobuffer Messaging

Before we invoke the tracker software from a higher level project, which is more practical than using an external batch, we will now introduce the protobuffer tracker code. In the Object Relations chapter we developed a `traccar.proto` file with a relational `Device` message. Now we will implement a tracker protocol to send this message to a tracking system.

Please go to the `jeets-tracker` directory to look at the code.

Besides using protobuffers the tracker needs a communication technology and we will add Netty to the tracker to make it the TCP communication Component. The idea of components is to create core business functionality based on the imported libraries.

Similar to the `org.traccar.protocol` package the tracker includes a `org.jeets.tracker.netty` package to handle different protocols in a `.proto` format. For a start we will add the classes `TraccarSender`, `TraccarMessageHandler` and `TraccarAckInitializer` to serve the `Traccar` protocol we have created.

The `TraccarSender` can send Traccar messages with the method

```
public static Acknowledge transmitTraccarObject
                (Object traccarObject, String host, int port)
```

What you see in the code is the standard Netty implementation to send binary information. There is no special treatment to send protobuffers since these can be created with the protobuffer protocol and library as demonstrated in the `jeets-protocols` tests. The `TraccarMessageHandler` actually implements the method to send the binary object via a Netty channel:

```
public Acknowledge sendTraccarObject(Object traccarObject) {
    channel.writeAndFlush(traccarObject);
    Acknowledge ackmsg = waitForAcknowledge();
    return ackmsg;
}
```

The `waitForAcknowledge()` method is a construct to keep the connection open until the message transfer is acknowledged by the server and the transaction is complete. Then the *client* has to decode the protobuffer `Ack` message from the server response. Therefor we have to create a receiving method to handle the incoming data. This is achieved by creating a `TraccarAckInitializer` with the usual Netty channel pipeline constructs.

Before you start to wonder about building a channel pipeline for Traccar messages you should look at the home page `netty.io` to find that Netty implies protobuffer handling! This combination of two JEE Components is implemented and underlines the book title to apply these combined components for fast implementation and best performance.

In the end decoding a protobuffer message is achieved in four lines

```
channelPipeline.addLast(new ProtobufVarint32FrameDecoder());
channelPipeline.addLast(
        new ProtobufDecoder(Acknowledge.getDefaultInstance()));
channelPipeline.addLast(new ProtobufVarint32LengthFieldPrepender());
channelPipeline.addLast(new ProtobufEncoder());
```

where you only have to supply the message type you want to decode, i.e. `Acknowledge.getDefaultInstance()` and then you simply add your business code handler

```
channelPipeline.addLast(new TraccarMessageHandler());
```

to finally receive an `Acknowledge` Java Object which can be handled by the programming logic.

If you compare this implementation to the complex Garmin implementation demonstrated in Section 2.3.2 you should feel the great relief of this approach!

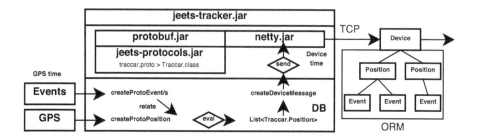

Figure 8.2 The `jeets-tracker` software provides a protobuffer protocol to create network entities from external data. The list (queue) of ORMs represent the database and Netty wraps the TCP communication to send as many ORMs as appropriate.

8.3   TRACKER ARCHITECTURE

The JEETS tracker is a simple software for many appliances. It could be a start to implement a real tracker in combination with an Android device that provides GSM and GPS access. For this book it will be used to send messages to the emerging JEE tracking system and support ongoing development. Figure 8.2 shows a simplified architecture that we will now transform into a construction plan in Java Language.

The process of modelling real world objects in software Objects (sometimes) is not as trivial as it may seem. If you think about the tracker settings by distance or time intervals mentioned earlier you will find that these rather belong to the device attached to the tracker. Some algorithm has to retrieve space and time coordinates from the actual GPS hardware unit as they become available. Therefore the JEETS tracker becomes much simpler and is basically stripped down to a TCP communicator. The use case for developers is as simple as:

```
Tracker tracker = new Tracker(host, port, uniqueId);
    :
// create builder and set protocol attibutes as appropriate
    :
tracker.sendPositionProto(positionBuilder);
```

By creating a **new Tracker** we define the server and port implying the protocol decoder and the **uniqueId** which should be registered in the GTS. The client code can **sendPositionProto**s as needed, i.e. by intervals, events or whatever logic. Note that every position is added with GPS coordinates, the timestamp reflects the actual *fixtime* of the GPS unit (with the related event time) and should never be modified. Internally the tracker will collect these positions in a **messageQueue** and then try to send them to the server until it is empty. Every time the tracker starts sending positions the *devicetime* is reset to the current time. Network problems or broken connections due to tunnels should not be raised to the surrounding software using the tracker – as a message buffer.

8.3.1 Implementation

Any 'device' can add, i.e. attach the `jeets-tracker` to its `pom.xml` file as a component. This implicitly provides the Traccar protocol with protobuffer messages which can be created and set by the device. We have defined the tracker method

```
tracker.sendPositionProto(positionBuilder);
```

as the major interface to the tracker. After creating a `ProtoPosition` with a GPS coordinate any number of attributes can be set until it is passed to the tracker.

Internally the method simply pushes the incoming position to a message queue and starts a message loop in a separate Thread to keep sending whenever possible until the queue is emptied. This behavior is very convenient for developers and the `Main.main` method provides the following practical use case:

Please *stop* your Traccar server, compile `jeets-tracker` and enter the (single) command line:

```
java -jar target\jeets-tracker-x.y-jar-with-dependencies.jar
     localhost 5200
```

You should see the tracker output in your console. The main method representing the external tracker environment creates a position with a GPS fixtime every ten seconds (time interval). As soon as the tracker receives a position it ensures that the `MessageLoop` is running and pushes the position into the `messageQueue`.

```
main: created list with 6 positions.
main: fixed 1. position at 05887   [ms]
Starting MessageLoop thread
DeviceBuilder filled with 1 positions.
'pb.device' send 1 Pos to 'localhost:5200' at 05896 (0 msgs queued)
Connection refused: localhost/127.0.0.1:5200
Transmission failed, trying again in 10 seconds
```

The `MessageLoop` running in it's own thread creates a new `DeviceBuilder` to wrap one or more positions in a single message. The protobuffer *builder* mechanism indicates that the message is still in the creation process and remains modifiable until it will actually be `.build()`. This is important as we want to set the device time for every position just before sending it.

Inside the `transmitTraccarDevice` method all positions are traversed to update the device time. If you recall the modeling of the proto messages we had decided to leave the device time in the position message in case we want to send single positions. Therefore the loop (duration) over all positions can provide slightly different times for each position. If you don't like that you could move the device time field into the device message and set it once for all positions – to save some additional processing time and network size.

Since the server is not running and the tracker can't get rid of the first position immediately the message loop keeps trying to connect in a separate thread. After the second position the 'device builder is filled with two positions:

```
main: fixed 2. position at 15889
DeviceBuilder filled with 2 positions.
'pb.device' send 2 Pos to 'localhost:5200' at 19245 (0 msgs queued)
```

We have learned that a device message can carry any number of positions. On the other hand we want to avoid large messages on the GTS port as it can slow down the complete network traffic of concurrent devices. In order to avoid large device messages with many positions the tracker offers the variable maxNrOfPositions to restrict the message size. After the device builder has reached this number of positions the newly arriving messages remain in the queue:

```
main: fixed 3. position at 1507480625890
DeviceBuilder filled with 2 positions.
'pb.device' send 2 Pos to 'localhost:5200' at 30272 (1 msgs queued)
```

Finally after all positions were added to the tracker the message loop keeps running and is waiting for the server to become available. This simulates a car driving through a tunnel or an area with bad coverage.

```
main: All device messages posted
DeviceBuilder filled with 2 positions.
'pb.device' send 2 Pos to 'localhost:5200' at 74468 (4 msgs queued)
```

Now you can start up your Traccar server and you can witness how the tracker transfers three device messages with two positions each – as soon as possible. Note that the tracker opens and closes the connection for each device message to avoid a longer session which would influence the other trackers communicating with the same server port.

8.3.2 Performance

In the context of SDC this client server communication can become very dynamic and the server can provide hints via the ACK message (that you can freely define for your purposes). In case the server detects high network traffic he can ask the trackers to send more device messages with fewer positions or vice versa. This can easily be implemented by adding a setter to the tracker:

```
tracker.setMaxNrOfPositions(10);
```

which can be applied any time.

In SDC scenarios it is also common practice to supply a `minNrOfPositions` for a single message. One reason for this is (country dependent) privacy protection which doesn't allow subsequent tracking for more than some minutes. Another good reason is the creation of a tiny route that can be analyzed on its journey through the system.

One of the great challenges of *competing* global players in the automotive business is to provide a single *common* Service reporting traffic situations live – for *all* self driving cars of different vendors. Submitting single positions from different anonymous vehicles is pretty useless. By submitting a `minNrOfPositions` for every vehicle we gain dynamic information. We can analyze and provide exact speed and course from position `n-1` to position `n`. And we can create some simple plausibility checks and throw a route away that could confuse the system logic. Plausibility can also be applied inside the driving car by comparing GPS and OBD data for example. From a GPS trace with three positions you can determine the exact 'GPS course' and compare it to the vehicles orientation provided by a gyro sensor. If these values deviate the vehicle is drifting and should raise a warning to other vehicles surrounding it. The 'traffic server' can then be fed with plausible motion data.

The decision between `min-` and `maxNrOfPositions` can also be implemented dynamically depending on the vehicle itself. If the vehicle approaches a complex traffic situation you should keep in mind that a vehicle travels through a city at 10 to 30 meters per second! In order to avoid a collision the client server communication should perform within a second to slow the vehicle down in time. Depending on the project you can calculate the manageable distance for a successful alarm measure...

The trackers main method can be used to get some performance figures – for your development environment with a local host connection. The Traccar position entity provides three relevant timestamps:

1. The *GPS fixtime* is the actual time when the event has happened and naturally is synchronized with the GPS coordinates to describe where it occurred. This GPS 'stamp' describes and event in the real world, before this event is handled by the technical components.

   The chronological order of GPS fixes describes the exact physical trajectory of the tracked object!

2. The *devicetime* describes the time when the tracker connected to the server to submit the collected events. In a real world scenario fix- and devicetime can be hours and days apart from each other.

   Just think of a test car collecting traces on Friday in an area with bad network coverage. At the end of the day the car is parked and the messages are still 'in the car'. The next working day (Monday) when the car is started the tracker connects to the server and 'flushes' the events from the previous recording to the server where they can be analyzed.

TABLE 8.1 Transmission Times [mm:ss.SSS]

DB ID	fixtime (10 sec)	devicetime (2 pos/msg)	servertime
1822	42:17.330	43:46.006	43:47.106
1821	42:27.332	43:46.006	43:47.095
1820	42:37.332	43:46.315	43:47.092
1818	42:47.333	43:46.315	43:47.011
1819	42:57.333	43:46.342	43:47.012
1823	43:07.334	43:46.342	43:47.108

3. The *servertime* depends on the server implementation. In case of Traccar the incoming position is first decoded with our `ProtobufferDevice-`
   `Decoder`, which immediately acknowledges the reception. Then the position entity travels up and down the channel pipeline and is enriched
   with an address by a geocoder, i.e. an external resource, before it reaches
   the `DataManager` to set the server time just before persisting it.

Let's look at the milliseconds of the complete process in Table 8.1. The
most important piece of data is the fixtime with coordinates of the real event.
The column reflects the 'time interval' of ten seconds. As you can see the milliseconds slightly deviate from the exact interval. The device column reflects
how the tracker has submitted three device messages, i.e. device sending times
with two positions each.

This has happened at `43:46.006` *after* the six positions were queued until
`43:07.334` when the Traccar server was launched to receive the messages. If
you want to analyze the route, display it on a map and so on you should
always refer to the fixtimes and the table is ordered accordingly. In the last
column we can see when the position has reached Traccar's `DataManager` for
persistence:

```
.setDate("serverTime", new Date())  // set Server Time
.executeUpdate());                  // sets Database ID
```

From the column Servertime you can actually tell the concurrency of receiving unordered positions inside the Netty pipelines. The ordinal database
ID shows the erratic order of writing and waiting for the database resource!
In case of high traffic these columns can appear in different order and even
longer times apart. Please be aware that the timing values refer to your local
PC via a local network and should only be seen as relative values. For a high
performance production platform they must be determined explicitly in order
to calculate the overall maximum performance – per port and hardware.

Again, the span from device- to servertime depends on the server implementation. If we would set the servertime inside the `ProtobufferDeviceDecoder`
directly before sending the `AKC` we would 'see' the network latency without
server- and database processing.

8.3.3 Binary Network Format and Size

The second performance aspect implied in the transmission times is the size of the messages traveling over the network between encoding by client and decoding by server. If you are using Traccar you can experiment by sending device messages with one, two, three .. positions and look at the log file for the actual hex message coming in[3]:

```
13:52:21 DEBUG: [26D9AA68: 5200 < 127.0.0.1]
HEX:650a0970622e646576696365122b08819bea87f02b10bedfe887f02b18012134de
4e09a5834840297605b209f63628403100000000000207740122b08819bea87f02b10d0
ade987f02b18012131815306ab834840293909e74b56382840310000000000f07740
```

These messages can be reduced by changing the switch inside the proto file

```
option optimize_for = SPEED;
```

from SPEED to CODE_SIZE. This way the messages become smaller, but (sequential) operations take longer. Depending on the size (and number of proto objects) of your messages you can test the system behavior. For very large messages you could even add a zip and unzip operations on both ends. We'll stick to SPEED for our intentionally short messages.

In the next section we will create a client server test for the tracker and use Netty logging to analyze the binaries

```
         +-------------------------------------------------+
         |  0  1  2  3  4  5  6  7  8  9  a  b  c  d  e  f |
+--------+-------------------------------------------------+----------------+
|00| 56 0a 06 74 65 73 74 49 64 12 4c 08 f6 cb b7 8f |V..testId.L.....|
|10| f0 2b 10 f6 cb b7 8f f0 2b 18 01 21 89 d2 ff ee |.+......+..!....|
|20| f4 83 48 40 29 16 6a 2c e5 a5 34 28 40 31 b2 9d |..H@).j,..4(@1..|
|30| ef a7 c6 d1 74 40 39 14 ae 47 e1 7a 14 d6 3f 41 |....t@9..G.z..?A|
|40| 1d 5a 64 3b df 07 59 40 49 71 3d 0a d7 a3 b0 28 |.Zd;..Y@Iq=....(|
|50| 40 5a 04 08 03 10 02                            |@Z.....         |
+--+----------------------------------------------------+----------------+
```

of a device message (with positions and events)

```
uniqueid: "testId"
   position {
      devicetime: 1507565889014
         :
      event {
         event: KEY_ALARM
            :
```

---

[3]divide the length of characters by two, since two characters display one byte.

8.4 TRACKER TESTING

We have created tests to create protobuffer messages in the protocol module. The tracker module adds Netty,i.e. TCP functionality to actually send these messages. In a first test case we don't want to include the Traccar server, since it would require an integration test to start the server, start the tracker to send a message and assert that it arrives at the server, which sends back an ACK.

Netty wouldn't be such a well accepted component, if the code could only be tested by integrating client- and server pipeline in one test case. In the `ProtobufferDeviceTest` you can see how to `setup()` and initialize a client and a server to simulate the protocol exchange within a single project. By using the `LocalServerChannel` and `LocalChannel` you can setup a local transport which allows in-VM communication.

Before we can use our tracker class `TraccarAckInitializer` for testing it needs to be improved by defining the channel type with generics

```
TraccarAckInitializer<T extends Channel>
```

in order to apply the initialization of the the protobuffer pipeline with a `SocketChannel` or the `LocalChannel`. Note that the `LocalChannel` does not require SSL context, host and port:

```
new TraccarAckInitializer<SocketChannel>(sslCtx, host, port)
new TraccarAckInitializer< LocalChannel>(  null, null, 0)
```

Due to the lack of a Server we have come up with some server code to receive, process and respond to the client message. With the symmetry of Netty code we can easily implement a simple `TraccarServerHandler` to receive the `Traccar.Device` message and create a `Traccar.Acknowledge` message in return. This handler is placed in

```
.childHandler(new ChannelInitializer<LocalChannel>() { ..
```

to setup the pipeline. The main difference to the `TraccarMessageHandler` is the Traccar message type

```
new ProtobufDecoder(Traccar.Device.getDefaultInstance()),
```

Here we can clearly see the advantage of using ORM messages for one port instead of using separate ports for positions and events, recomposing their relations on the server side. And with the `TraccarServerHandler` we have written some initial server code we can pick up later.

8.5   CONCLUSION

What you get with the `jeets-tracker` is a seed component to create your own tracker logic. The code is kept to a minimum and you can experiment with the tracker while switching the server on and off. In the next chapter we will run tracks that last more than 30 minutes to experience the immanently *unreliable* and *unstable* client server connection.

We have carefully modeled the `Device`, `Position` and `Event` messages into a single object relational message with respect to the Traccar system. For a larger set of messages the Garmin FMI message catalog was listed to provide a picture. For SDC systems we need much more messages and even more for SDC development systems. Yet the modelling with `.proto` files is the same and you can start by creating complex files and then adopt the software to handle them in the same manner.

For a complex situation where the autonomous car software has to make a decision to give way for a pedestrian crossing the road you can record about 8 GBit per second, one terabyte of data in 15 minutes! The automotive industry offers measurement platforms for data acquisition with high bandwidth, high resolution sensors etc. to capture (raw) data. Synchronization is specified by IEEE802.1AS with hardware time stamp. In order to extract information from the data flow use cases are defined with especially designed message catalogs. This book describes the bits and pieces of device communication with Netty and protobuffers and it should be easy for you to create your own set of messages from there on.

The JeeTS Player

CONTENTS

9.1 INTRODUCTION

The major difference between Standard and Enterprise Java software is that the latter is created with – preferably standardized, dedicated and most of all reusable – components to form a software and add pure business code. Therefore we have defined the tracker as an isolated Component 'to equip something with'. The `jeets-tracker` is ready to be placed in a vehicle and hooked up with its services to retrieve and send live information.

In the last chapter we created the tracker with a protocol to send Traccar messages. The tracker software can not acquire a GPS coordinate and needs to be 'fed' with valid messages to be sent. Imagine you want to place the `jeets-tracker.jar` in some client, which provides a Java VM, like the Traccar Android client. Then you can use this environment to receive GPS coordinates, analyze, optimize them and 'fill' a device ORM message to send it to the GTS.

Another important scientific fact to justify on board processing is that *sampling*, i.e. sending erratic positions and events, is not the same as *measuring* the exact vehicle behavior. With one position before and another behind a curve it is *not* possible to detect the exact zenith of the trajectory. The Controller should provide plausible results (information) rather than raw probes (data).

Besides evaluating and collecting GPS positions continuously the software can start a GPRS process to create a TCP connection when the internal database, i.e. message queue is filled (to a predefined threshold). The ACK

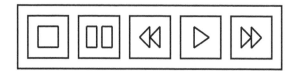

Figure 9.1 The `jeets-player` software provides a simple way to play, replay, fast forward an GPS track similar to the functionality of a of music player panel.

can verify the successful transmission or the NAK can restart the complete process.

An important guideline for the complete JEETS development is that it does not propagate anything 'into' a vehicle. It will rather provide a module to exchange important information between client (vehicle) and server. Even if it would raise a warning like `'traffic jam 500 meters ahead'` it would be up to the Original Equipment Manufacturers (OEMs), i.e. the car industry (hardware), to propagate this to the vehicle and slow it down.

9.2   GPS PLAYER

The main idea of the `jeets-player` is to re/play the GPS track of a single device, vehicle. A GPS player is a very helpful tool at development time. Without it every developer would need a hardware tracker which requires complex handling or even worse, someone to drive around with it – during working hours. With a player software you can grab one of thousands of available tracks, replay it and feed the positions into our `jeets-tracker`. Beyond that a number of simulations can be used to *replay* complex traffic situations. The term player expresses a playback application for previously recorded tracks as indicated in Figure 9.1. The idea is to specify a source of a recorded GPS track or to generate one as needed.

9.3   DEVELOPMENT CYCLE

In Section 2.2.1 we created an `EchoClient` to support development by sending tracking messages to the server as needed. In the Traccar Android client we can access the location API as described on page 88. Now we are creating a player to create messages and send them.

In this early development phase we want to avoid deploying software on a client hardware and then having to move the device around to receive tracking messages. We would like to work on the complete JEETS repository and be able to simulate GPS traces *while* implementing the logic. This sounds more challenging than it actually is and we are well prepared.

Development with client and server on a single computer and communicating via `localhost` can also be used to determine exact processing times

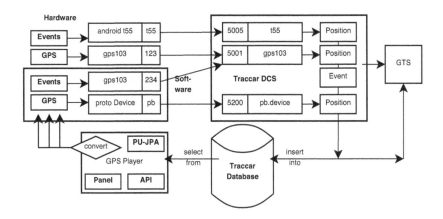

Figure 9.2 The `jeets-player` can be used to replay the tracks recorded by any tracking device and can implement any protocol *Encoder* from Traccar entities defined by the Traccar persistence unit.

as demonstrated earlier. Both Components access the identical time source and by looking at the timestamps in the database you can get figures for fix-, device- and server times.

In the `jeets-tracker` we have created a `Main.main` method to run via command line. There we use the method `createSampleTrack()` to create a tiny track with a few positions to send. We have programmed the method to send the positions in ten second intervals which allows some observations during the route. Anyway the route is not plausible with this timing – its merely a technical helper.

By creating a separate `jeets-player` module we can replace the trackers `main` method with a realistic playback of a track recorded in reality (at a certain time of day, year, season, city, etc.). For the readers of this book it should be easy to record tracks with Traccar and at the end of this chapter they can be replayed and observed on the Traccar frontend.

To access the Traccar database we have already created a persistence unit and by placing this PU in the client environment we can access the database to *retrieve* GPS traces which we can collect every day. By creating the JEETS player project we finally have a feasible setup for developers to establish the cycle and use the database as our 'GPS unit' as shown in Figure 9.2.

9.4 IMPLEMENTATION

The complete player implementation is based on a single list of entities – not protos! Entities represent the final data format in the system and are familiar to more developers than the binary network formats. Any Java developer can create a `List<Position>` and pass it to the player.

Another aspect is the reusability of the player *inside* the server environment as it is specified by JPA. We could use the player to play the track of an Enterprise Java bean!

The construction of a valid player is achieved by

```
Player player = new Player(positionEntities);
```

at construction time. Or you can add or change the position list directly and later, when the list playback has ended you can set another list.

```
player.setPositionEntities(positionEntities);
```

To receive each position with the exact time delay as in the original track the player project provides the `PlaybackListener` with the method `receivePositionEntity(Position positionEntity)`. After adding this method to any `client`[1] you may create you can start the player and process the positions as needed. For example you can pass them to a tracker to send them:

```
Tracker tracker = new Tracker("localhost", 5200, "pb.device");
Client receiver = new Client( tracker );
player.addListener(receiver);
player.startPlayback();
```

Internally the player works similarly to the tracker by invoking a `PlaybackThread` to traverse the position list. While the tracker 'gets rid' of its positions by sending them, the player's list is not modified at all. The player is a tool to playback the complete list, i.e. track or portions of it. This way we can develop a solid API to operate on a single Java collection.

You should carefully study the Maven design of the `pom.xml` files for PU, protocol, tracker and player. The tracker naturally has to 'know' the protocol and the system model in order to convert them. Therefore the protocol project provides the `Transformer` which can be applied to the received position entities.

If you plan to use the tracker to send Traccar messages for *any* tracking protocol supported by Traccar you can add the counter parts of the protocol *decoders*, the protocol *encoders* to the tracker step by step.

9.4.1 Player API

As indicated in Figure 9.1 the player can be compared to a music player. After loading the 'playlist' we can apply the player's panel to

```
.startPlayback()
```

```
.startPlayback(Date from, Date to)
```

```
.pausePlayback()
```

---

[1]The player's main method demonstrates how to send messages to a `SampleReceiver`.

```
.continuePlayback()

.stopPlayback()
```

Note that the `jeets-player` provides a simple implementation which can be used 'out of the box'. You might come up with some more ideas to support your business case and can easily add useful methods to:

start playback according to weekday and daytime

ignore weekday, use daytimes only

repeat a track n times

etc.

Another feature found in track simulators – the player is only one of them – is the adjustment of a speed factor. It's good to have a trace from a trip over 50 miles and one hour duration. While the SDC production system has to process many cars concurrently the developer might be interested only in this one trip. Then it should be no problem to speed up the playback in order to replay the trip in a few minutes – not a real difference for a modern PC.

```
player.setTimeFactor( 10 )
```

9.4.2   Plausibility

The player is introduced here for the upcoming development of tracking server software. But not all player features should not be used to send tracks to the production system! A production system should always receive real world data live and a player with double, triple .. speed will probably raise plausibility problems which can cause a lot of pain for a server – unless it applies the same time factor.

Another plausibility issue is the order of the recorded track. For a programmer it is easy to traverse a track backwards. Think of the implications! The 'vehicle' would travel the wrong way against the traffic. A bicycle would climb a hill at world record speed, etc.

As a conclusion we should always use the player with care and with respect to the targeted system. And after all a recorded track is never really live – with one exception! Public transit is carefully planned into the future and we will access this kind of data in a sample project in Chapter 11.

9.4.3   Usage

First you need to choose and configure a database to retrieve tracks for playback (input) by overriding persistence.xml

Then you should determine an interesting track and define the times from, to and a unique ID. You can do this with plain SQL queries

Finally you place the three parameters in the Java code which creates the query

Now you can start the player

9.5 SIMULATION

The `jeets-player` is a simple, yet very useful application and it should serve as a starting point for your own implementations. For a developer of the car's internal components the player can easily be recoded to a `jeets-simulator`.

The simulator does not simply replay a track, it rather 'listens' to its traffic environment and can react if needed. The Simulator API should work similarly to driving a car (see [19]):

```
.move(direction, speed)
```

```
.turn(direction) etc.
```

my JEeTS Client

CONTENTS

10.1   MAVEN ARCHETYPING

This chapter wraps up the client development. Do you recall every detail of the protocol, persistence unit, tracker and player's design processes? As you move on to the system you don't even want to remember and free your mind for new challenges. In a large enterprise you can have one team for device communication and you can always ask these colleagues *in case* you do want to know the details. Then again communication (between people) is the most expensive (sometimes missleading and confusing) process of all!

It would be laborious to teach every system developer how tracker and player can be setup to collaborate. He probably wouldn't care much about binary message formats – he wants to know how to receive system messages, i.e. entities in his environment.

For this very common situation we will introduce Maven's archetypes [24] as *the* solution. A Maven archetype is a project developed just like any other – only it can be `generate`d into a new project structure. We will start the `my-jeets-client` project just like the earlier projects we have realized. The `my-` in the project name should indicate that this is a complete software that `you` can `generate`. Then you can immediately run this software and place it in your environment. After successful tests you can modify the software according to your needs without modifying the `my-jeets-client` project source. Maven archetypes are great to standardize new projects in an enterprise based on proven and frequently used company modules.

Obviously each supplier company, each car vendor creates his own 'client box', even for dedicated car types, with complex electronics and software to prepare and *receive* data for processing from dedicated server applications and services. Since these projects are developed under non disclosure this book can only provide *a* client environment. We will focus on some nice scenarios more concerned with enriching tracking server functionality. The important message is that any enterprise developer can generate a client with the latest software modules.

On the other hand the developer should keep an eye on subsequent releases with more functionality. It is his responsibility to manage the code inside the new project and be prepared to move it to yet another client software with new features, like an ftp download for fresh maps. We will look at this use case in the next chapter where we will create yet another Maven project which can be used to replace the generated default implementation. After a client update only the artifact needs to be placed in the pom file.

The original client project will be setup to replay any track to a server. In an commercial SDC project the Client becomes more and more complex. For example the software can have access to an onboard digital map which is a great reliever of server processing. As a matter of fact many OEM projects don't transmit GPS coordinates at all. They snap the GPS coordinates to map links and transmit the link ID, driving direction and offset distance from the beginning of the link.

Therefore you can imagine additional software to hold the digital map, update it from an ftp server according to your driving environment, to do the map matching and finally sending map matched information instead of geo coordinates. The client software, kept in the repos `jeets-clients` folder, can increase just like any other Maven project and you could add the map matching. As a next step a receiver of TMC messages can be implemented etc.

At a certain level of development the client software turns into the actual software installed in the car. Then of course the player as a virtual device is replaced by the car as the actual device and the project file could add modules provided by the automotive companies' developers of embedded software:

```
<artifactId>audi-a4-canbus</artifactId>
    :
<artifactId>here-eu-hd-map</artifactId>
    :
```

10.2   IMPLEMENTATION

The construction of a player was prepared on page 104 in the `Main.main` method to wire player and tracker together in the `my-jeets-client` environment:

```
Player player = new Player(positionEntities);
Tracker tracker = new Tracker("host", 5200, "pb.device.echo");
MyClientDevice receiver = new MyClientDevice( tracker );
player.addListener(receiver);
player.startPlayback();
```

We have carefully separated the designs of player and tracker and now we combine them in a new project. The developer only has to know system entities and the tracker will take care of transforming and transmitting the protobuffers. The developer only has to create a `List<org.jeets.model.traccar.jpa.Position>`. The client project provides a simple example for retrieving positions from the tracking database as is shown in Figure 9.2 on page 103 to set up his development cycle.

As the client project is *not* a reusable Java component, but rather a starting point for a new software. The database and query parameters won't work out of the box. After generating a new project, which we will do in the next chapter, these parameters should be adapted to your environment. You can use plain SQL to identify the tracks you want to replay to the system.

```
private List<Position> selectPositionsFromDB() {
    String jdbcUrl = "jdbc:postgresql://localhost:5432/traccar3.14";
    String fromDevice = "HHA-U:U1_HHA-U";
    Date fromDate = parseDate("2017-06-05 13:09:00");
    Date   toDate = parseDate("2017-06-08 16:36:00");
                :
```

This approach is the simplest and most standard way to apply the persistence unit we have created for the system and was therefore chosen as a good starting point. What you get with the `MyClientDevice` class is a skeleton for a new software to be deployed on a client hardware!

Once you have a running software and have successfully transmitted positions to the server, as a proof of concept, you can start developing. For one thing you can create your own `List<Position>` as we will see in the next chapter. Note that the method

```
@Override
public void receivePositionEntity(Position positionEntity) {
```

actually represents the information gathered from a GPS unit and internal attributes from the device – which can be a car. Therefore this method can be removed during ongoing development and replaced by a method to gather values from an actual hardware. Or you can coordinate both methods to switch from JEETS player to hardware. Once the software is stable the player, being a development support component, can be removed and should not appear in a production release anymore.

That's all there is to it!

Within the context of this book, this project may seem to tiny too draw much attention. Then again we have pointed out that 'customers' generating their new project from the archetype will experience a fast learning curve to combine all client components – without caring much about implementation details.

Let's go through the process and generate some traffic
with a new client software in the next chapter.

JeeTS GTFS Factory

CONTENTS

11.1 TRAFFIC

We have created the player to playback tracks and indicated that the player software can be recoded to a simulator and there is more. If one player provides the track of a single vehicle then you can create a new client software to replay many vehicles and actually generate traffic on the server side!

If you don't have access to GPS traces you should use your smartphone to track your daily trips to and from work with Traccar. As an example we'll assume that you have a strict nine to five job. With recordings of five to ten days you can create SQL statements to select your tracker and daytimes from 8:30 to 9:00 and from 17:00 to 18:00[1]. Then you create two players for every day, start all of them simultaneously and watch where they approach each other in opposing directions.

Depending on what you actually want to see on the map you could fine

---

[1]trips after work usually don't follow the shortest way home!

tune the SQL statements with precise timestamps in a way that all tracks simultaneously arrive at a certain intersection or small area. And you could add the player's panel as indicated in Figure 9.1 on page 102 to the client environment to implicitly operate on all players simultaneously.

With access to real recorded traffic not only the daytimes matter. Traffic prediction is a science of its own and yet traffic can only be roughly predicted by taking weekdays, holidays, season, construction work and many other factors into account.

This book indicates how to apply connected car technology to 'traditional' GPS tracking systems. As fleet management is the most commercial use of tracking systems a client could be coded to replay the fleet's motion and again serve development towards a smarter system. Notifications can be raised to the logistic system, if a colleague driver is close to another truck and could direct them to a meeting point to exchange merchandise.

In the next part of the book we will go through the development of Jᴇᴇ server components and it might be a good exercise at this point for you to setup a client from the examples above.

In the end you should be prepared to simply enter a command line to start your traffic service and develop the server components according to your own mission!

11.2   GENERATE MAVEN ARCHETYPE

In order to create a traffic application we can now use the Maven archetype project we created in the last chapter. In the JᴇᴇTS repository you will find a project **jeets-gtfs-factory** which was created this way and now we'll go through the creation process. You can follow the instructions by creating the same project with your package names etc.

Before you can apply the archetype you need to run Maven install on all of its components to place the archetype in your local repo. In short: run Maven install for the complete JᴇᴇTS repository (or at least all steps including the archetype project). Please create a new directory anywhere on your PC, for example **my-gtfs-factory**, and enter this directory. Now you simply enter the command line

```
mvn archetype:generate                                    \
    -DarchetypeGroupId=<archetype-groupId>                \
    -DarchetypeArtifactId=<archetype-artifactId>          \
    -DarchetypeVersion=<archetype-version>                \
    -DgroupId=<my.groupid>                                 \
    -DartifactId=<my-artifactId>
```

and watch how the new project is generated.

Now you can test the new software by adjusting the SQL statements with the database context you have generated with Traccar.

Next you should open the pom file and add the artefact jeets-gtfs-factory and run Maven install.

Before you can start coding in your new project we will introduce GTFS and the jeets-pu-gtfs project.

11.3 TRANSIT FEEDS

Today in a world of growing transparency more an more transit agencies provide their schedules as downloads. When Google Maps turned into one of their major products many efforts were made to standardize transit data to incorporate public transportation.

GTFS - General Transit Feed Specification [25]

> "The publicly and freely available format specification, as well as the availability of GTFS schedules, quickly made developers base their transit-related software on the format. This resulted in 'hundreds of useful and popular transit applications' as well as catalogues listing available GTFS feeds. Due to the common data format those applications adhere to, solutions do not need to be custom-tailored to one transit operator, but can easily be extended to any region where a GTFS feed is available."

If you think about it, public transportation is highly predictable and by playing the schedules live we get a nice 'grid' for time and space. We don't want to miss this opportunity to create a 'public transportation factory' that we can use on server side to analyze traffic.

In order to follow the implementation you should go through Appendix D to setup a Postgres database with GTFS data of a place you know very well.

11.4 GTFS PERSISTENCE UNIT

After creating a GTFS database the natural next step in JEE is to create a GTFS persistence unit just like the Traccar Persistence unit we have created to get hold of the system model via JPA specification.

...you can find the GTFS persistence unit in the `jeets-pu-gtfs` project. Note that this data model is based on the GTFS import with the `OpenTransitTools/gftsdb` tool described in Appendix D and *does not* provide any table relations ...

Relations can be modeled in the object relational model, i.e. coded in the GTFS entities (test retrieval of children).

11.5 TRANSIT TERMINOLOGY

A persistence unit is more than a technical construct, entities are the main actors of / in a system and highly support communication about the domain. We can look at the GTFS model with the common sense of a passenger, i.e. in his domain language:

First of all transit services of an *Agency* are based on stops where passengers can enter and exit transit vehicles. The city of Hamburg has about 24.000 stops and these *stops* form the base map for public transportation.

A 'chain' of stops served by a single transit vehicle is specified as a *route* in GTFS [26]: "A route is a group of trips that are displayed to riders as a single service." Another maybe more familiar word for a route is a line.

Once a passenger has planned to travel from one stop to another he needs to determine one or more routes that connect these stops. And then he has to decide when to depart from a stop in order to reach a destination stop in time. GTFS provides *stop times* for arrival and departure at each stop. Hamburg provides 1.4 million stop times per quarter of a year! Stop times are related to a *calendar* and *calendar dates* specifying the valid / included and invalid / excluded dates (weekday, weekend, holiday, special events, etc.).

When boarding a transit vehicle the passenger is going on a *trip*. Each route (spatial) offers a number of trips (temporal) and has a *shape* defined by (lat,lon) waypoints.

A GTFS feed can be described as follows: [27]

"A GTFS feed has one or more routes. Each route (`routes.txt`) has one or more trips (`trips.txt`). Each trip visits a series of stops (`stops.txt`) at specified times (`stop_times.txt`). Trips and stop times only contain *time of day* information; the calendar is used to determine on which *days* a given trip runs (`calendar.txt` and `calendar_dates.txt`)."

11.6 TRANSIT FACTORY

We will now explore how we can apply the `jeets-pu-gtfs` project in conjunction with the GTFS database to create a live vehicle track. If you are new to GTFS you should gain some experience by first interactively using a SQL tool, like PG Admin, to identify transit vehicles leaving at the next stop from your home to another stop you know well. Once you are familiar with the data structure and relations you remodel your prepared SQL statements to 'JPA Java'. In the `jeets-gtfs-factory` you will find the `TransitFactory` class, which implements a top down search from GTFS data.

Let's go through the major entities / tables of the GTFS / db. For prototyping it is helpful to start by creating SQL statements and combining with named queries. In a next step SQL can be remodeled with entities.

11.6.1 route_type

When you want to explore the public transportation of a big city like Hamburg the `route_type` is a good starting point to filter a manageable number of vehicles. Have a look at the small table to find all `route_types` offered by the agency. The `route_type` defines the type of vehicle or transport and therefore implies the speed and distances covered.

route_type table !?

For Hamburg the `route_type` = 1 you get
"Subway, Metro – Any underground rail system within a metropolitan area" is a good start for visitors to cover the complete metropolitan area.

Now we can identify the Routes covered by `route_type` = 1 with

```
select * from routes where route_type = 1
```

to identify only

four U-Bahn: `U1`, `U2`, `U3`, `U4` and

six S-Bahn: `S1/11`, `S2/21`, `S3/31`

routes for the complete city.

11.6.2 routes

We'll pick the U1

```
select * from routes where route_short_name = 'U1'
```

to get: "Norderstedt Mitte > Hauptbahnhof Sued > Ohlstedt / Grohansdorf"

With the Route selection we have identified the `route_id` and the `agency_id`.

11.6.3 route_stops

With the `route_id` we can look up its stops. Note that the `OpenTransitTools/-gftsdb` loader additionally creates the useful `route_stops` table, which might be missing in other import tools. From this table you can retrieve an ordered list of stops in both directions.

```
select * from route_stops where route_id = '1466_1'
```

On the other hand not every trip must serve every stop. Another great helper for transit data is the agency's website where you can lookup any trip or schedule any time. If you'll have a look at `www.hvv.de` and click on 'Verkehrsnetzplan' you can find a map to visualize routes.

Below the map window you can uncheck busses, boats etc. and check U-Bahn to visualize the 'blue line' as the 'U1'. You can clearly see that the 'U1' is not a single line, but it actually branches off and each of the two branches is served at different times.

The `route_stops` table holds every stop but not every trip starts at the outer end. Therefore we will take a different approach: We'll pick two stops from the 'U1' somewhere around 'downtown' to get a good coverage of trips. This process could be automated with a thorough analysis of each line, but here we'll manually pick 'Farmsen' and 'Fuhlsbüttel'.

11.6.4 stops

Now we can join the `stops` and `route_stops` tables to lookup the stops with names and coordinates:

```
select st.stop_id, st.stop_name,
       st.stop_lat, st.stop_lon, st.parent_station,
       rs.direction_id stop_direction
  from stops st, route_stops rs
 where rs.stop_id = st.stop_id
   and st.location_type = 0
   and st.stop_name = 'Farmsen'
   and route_id = '1466_1'
```

The result may be surprising as it returns more than one stop for one name. To understand this another term, the *station*, is introduced in GTFS. The station 'Farmsen' is familiar to Hamburg's population, but you wouldn't make an appointment to meet at 'Farmsen' on the 'inbound' platform. The station is a collection of (different) transit vehicle stops like in/out bound subway / bus etc. and it is listed in GTFS data as the `parent_station`.

In order to find a trip from 'Farmsen' to 'Fuhlsbüttel' we could take a route's `direction` into account. A direction can be problematic without local knowledge just like there is no left or right in geography. For different digital maps you have to look up the definition of a direction and the rules to detect a car's driving direction.

We will use the above query anyway and simply use all returned `stop_ids` for the next SQL statement. By collecting all stops for the from- and to station we can find a Trip without using the `direction`. The query is wrapped in the method `findStationsInRoute(fromStop, routeId)`.

11.6.5 agency

It is generally mandatory to look up the agency for every route.

For one thing you may only publish transit data in conjunction with a weblink to the agency, i.e. `www.hvv.de`. Beyond this legal issue the agency provides the language to address local customers with an UI and most of all the agency provides the time zone for the routes.

If you develop software for your local agency sitting at your local PC you can easily oversee the timezone. If you are developing with Hamburg data while sitting at your PC in the US you would like to retrieve the vehicles running 'now'[2]. And last but not least you can only develop *one* global transit application by using the time zones for trips between these.

After setting your local PC time you have to convert it to the agency's time zone:

```
ZoneId currentZone = ZoneId.of(agency.getAgencyTimezone());
ZonedDateTime zonedDateTime
            = ZonedDateTime.ofInstant(departAt, currentZone);
```

11.6.6 calendar and calendar_dates

Before we can search for a trip with our parameters we have to filter the trips for a given day. Remember that stop times do not have a date!

In the method `getServicesForDay(zonedDateTime)` you can find the query on `calendar` and `calendar_dates` to retrieve the `service_ids` for a day from a calendar.

The `calendar` table provides the days for running trips while the `calendar_dates` table can add or subtract single days, i.e. override the calendar days.

11.6.7 trips

At this point we have gathered all parameters to finally search for the trip on the selected route, from one stop to another at a certain zoned day and time. Have a look at the Java method implementation

```
GtfsTrip trip = gtfs.getNextTrip
        (selectedRoute, fromStops, toStops, zonedDateTime);
```

to find the details. The search is executed in two steps:

Determine the services for the day
via tables `calendar` and `calendar_dates`.

Determine the trip
via tables `route_trips` and `stop_times`.

To differentiate start- and end stop the table `stop_times` is joined two times for departure and arrival which are distinguished with `dep.pickup_type = 0` and `arr.drop_off_type = 0`. Note that the SQL result is limited to one dataset (`limit = 1`) and will therefor always return the 'next' trip only.

Now we've found our trip and can work with the `GtfsTrip` entity. From this entity we can finally create the `List<Position>` that we want to run with the JEETS player.

---

[2]`Instant depart = new Date().toInstant() // local PC time`

11.6.8 shapes

Once we have found the trip with station coordinates and times we can feed it into the List<Position>. Then we can run a GPS track by transmitting the stops every x minutes. At this point we have no information about the trip between the stations.

Depending on the agency the GTFS distribution can include the full geometry of a route that can be reduced to a trip. In the case of Hamburg the data contains about 2.5 million shape points. And Hamburg has another nice feature: the route geometry is published and maintained on OpenStreetMap!

By adding the full geometry to our trip and setting Traccar to OpenStreetMap you can actually see the train transmitting a position with every change of direction to reflect the exact shape visualized on the map frontend.

With the route_id we can look up its geometry and the SQL statement

```
select * from shapes
  where shape_id = '23'
  order by shape_pt_sequence
```

is wrapped in the method

```
List<GtfsShape> tripShapes = gtfs.getShapes(trip)
```

Now we have all the information to create a large List<Position>; in the case of the U1 we have 1173 shape points for a 45 minute trip! This is a good timespan for development with a transmission every two or three seconds!

11.6.9 Transit API

By playing with SQL statements we can create an API (the TransitFactory) and finally we can use the Factory to create transit vehicles as needed:

```
getNextTrack("U1", "Farmsen", "Fuhlsb\"uttel", "2017-11-03T18:08:00Z")
```

From here each developer can create his own transit traffic by creating a trip for every route

```
getNextTrack("U1", "Farmsen", "Fuhlsbuettel", [now])
getNextTrack("U1", "Fuhlsbuettel", "Farmsen", [now]) // opposite dir
getNextTrack("U2", "FromX", "ToY", [now])
getNextTrack("U2", "FromY", "ToX", [now])               // opposite dir
  etc.
```

And this factory can be applied for every new trip starting next in an infinite loop.

IV

Enterprise Integration

CHAPTER 12

Enterprise Software

CONTENTS

12.1   STATUS AND OUTLOOK

Let us review what we have done so far – client software and protocol – and see where we're heading – server software and system model.

We have modelled data formats for the JEE tracking software to be created. By doing so we define *every* piece of data entering the 'system'. On the other hand it is impossible to foresee every detail that we might run into when the software will be created and organically grow. This dilemma has always been there: software is alive! Once you have reached the end you can start resuming from the beginning... therefore solid software is developed in cycles. We have picked the self driving car as *the* project changing the world as we know it and challenging the software development process.

Continuous integration, -deployment and -delivery describe an automated software release chain to propagate every little bugfix directly to the production system. Especially SDC scenarios are dealing with human lives and we already know before hand that we can not design the perfect system – on paper.

Therefore we will use *one* Maven Repository for the ongoing JEETS development to host *all* interdependent JEETS Components. After every change we can run `mvn clean install` to compile the sources in order of rising dependencies, run tests for every single component including the more complex Components relying on previously compiled components. Later we can add integration tests to simulate client server exchanges, etc. Maven is responsible for retrieving the JEE components defined in the project object models. The complete JEETS software is living in a single repository, which itself can become one of many other software in a continuous process.

Besides creating a relational data model we have provided a practical approach to modify the model and recompile the system against it. The build process will and *should* definitely fail (fast) if we remove the `latitude` attribute from the position entity. The protocol defined in the protobuffer `*.proto` file is initially reduced to a minimum of one single `Device` message. In the prototyping process this a good place holder for further implementations. The tracker implementation with only four lines of code (listed on page 91) is a good template for any other protocol. Adding fields and relations to the `Device` object does not require any change in the tracker. Later it will be easy for the developers to add more protocols – to the complete software stack.

As a guideline we'll just keep in mind that the Maven BUILD SUCCESS of the complete repository is the prerequisite for continuous integration – towards your system, which is currently represented by a running Traccar GTS.

Please go through Appendix C
to build the JEETS repository and import it into Eclipse

12.1.1   Design Constraints

One of the seldom mentioned constraints in IT business is that system design always has to respect the production system, which is generating money. Whenever you need to add new features (requested by the marketing team) you have to analyze the impact on the system architecture and make a decision to alter the software (and introduce some workarounds) or to create a new Component that can interact with the current system.

We have decided to create some new components, but we can not create a complete new system in the course of a book. Traccar is a well accepted GTS that we don't want to replace. We want to look at alternative implementations for some Traccar components and they should run in conjunction with Traccar.

In Chapter 7.5 we created a proto decoder that can be added to the Traccar system – with code modifications. These modifications have to be applied after every Traccar update – occurring frequently. The next components to be developed are a DCS- and an ETL component to run stand alone from the command line and persist collected messages in the Traccar database.

This approach could be called 'one way integration towards Traccar'. With this collaboration via database we can still use Traccar's administration (frontend) to register devices, assign them to people and groups. We can also use the map frontend to visualize the collected data and the reporting to aggregate data for accounting reports, etc.

Some years ago OpenGTS was the only Java open source tracking system with low level TCP implementations for the DC servers. When Traccar development started many small tracking businesses replaced the OpenGTS

DCS jars with the Traccar DCS which operates with the Netty framework. By treating Traccar DCS and OpenGTS as one system the transition was hidden to the customers.

12.2   ENTERPRISE INTEGRATION PATTERNS

Continuous integration (CI) aims at enterprise integration with the term enterprise referring to many developers, different teams, wide or globally spread (proprietary) systems of large companies. For example most globally operating companies provide LDAP[1] authorization servers to provide the identity of every single employee, customer and even visitor. This service can be used to open doors, to login to the personal workstation and so forth. Each authorization can be withdrawn with a single mouse click.

With CI every software can be equipped with credentials and then be tested against the real enterprise LDAP mechanism. In other words CI makes sure that the software not only works in an isolated setup, but adheres to the software of other departments, which can also be commercial products: Email, calendar synchronization, inventory, customer relations... any confidential software. This kind of integration has to deal with various problems before it can be released. Network connections, file systems, databases have to be accessible in the process and are constantly changing.

All of these challenges have led to Enterprise Integration Patterns (EIPs) and you can find a great introduction at [30] which supplements the book by Gregor Hohpe and Bobby Woolf [31]. We will not reproduce the content here. Instead we will apply one pattern at a time as we move on.

The title of this book actually implies enterprise integration. Up to this section we have looked at some JEE components like Hibernate in order to create a JEETS persistence unit specified by JEE JPA for database interactions. The course of the book reflects global software development techniques where these components become smaller and much more robust to operate in various environments.

To demonstrate this development in practice we have chosen Traccar as a single monolithic software and to reassemble a tracking system with individual standardized components which can be used almost any tracking software you may have in mind. Or to provide 'connectors' to feed existing systems with GPS information.

---

[1]Lightweight Directory Access Protocol

12.3   MONOLITHIC TRACCAR ARCHITECTURE

Let's have another look at the Traccar architecture and identify the major processing components. From there we can define some isolated JEETS components with similar functionalities. Starting in the `Main.main` method we can see how the `Context` is initialized with the `.xml` configuration. We find the the the creation of many `*Managers`[2]:

```
public final class Context {
    public static void init(String[] arguments) throws Exception {
        config = new Config();
        config.load(arguments[0]);
             dataManager = new DataManager(config);
           deviceManager = new DeviceManager(dataManager);
         identityManager = deviceManager;
      permissionsManager = new PermissionsManager(dataManager);
       connectionManager = new ConnectionManager();
         geofenceManager = new GeofenceManager(dataManager);
         calendarManager = new CalendarManager(dataManager);
     notificationManager = new NotificationManager(dataManager);
           serverManager = new ServerManager();
          aliasesManager = new AliasesManager(dataManager);
       statisticsManager = new StatisticsManager();
```

All of these managers belong to the `org.traccar.database` package and depend on the external resource 'database', i.e. `dataManager` – except the `ServerManager` which is responsible for networking and device communication. The `ServerManager` was already introduced earlier and is in charge of setting up the `BasePipelines` with all kinds of `Handlers`:

```
public BasePipelineFactory(TrackerServer server, String protocol) {
    // first decode binary protocol to system entity ..
                 this.server = server;
    // .. then handle entity in system ..
          filterHandler = new FilterHandler();
     coordinatesHandler = new CoordinatesHandler();
        geocoderHandler = new GeocoderHandler(..);
     geolocationHandler = new GeolocationHandler(..);
        distanceHandler = new DistanceHandler();
      hemisphereHandler = new HemisphereHandler();
  copyAttributesHandler = new CopyAttributesHandler();
commandResultEventHandler = new CommandResultEventHandler();
     overspeedEventHandler = new OverspeedEventHandler();
        motionEventHandler = new MotionEventHandler();
      geofenceEventHandler = new GeofenceEventHandler();
         alertEventHandler = new AlertEventHandler();
      ignitionEventHandler = new IgnitionEventHandler();
   maintenanceEventHandler = new MaintenanceEventHandler();
```

We have already highlighted that every protocol process starts with a `*Decoder` to provide the system entity `Position`. Then this entity is passed up the stream to go through different handlers

---

[2]depending on the configuration

```
public abstract class BaseDataHandler extends OneToOneDecoder {
   @Override
   protected final Object decode(ChannelHandlerContext ctx,
              Channel channel, Object msg) throws Exception {
      if (msg instanceof Position) {
         return handlePosition((Position) msg);
      }
   return msg;
   }
   protected abstract Position handlePosition(Position position);
}
```

to do something with the incoming position. And the second process is represented by the `BaseEventHandler` to create `Events` which are required for several use cases the system should serve.

```
public abstract class BaseEventHandler extends BaseDataHandler {
   @Override
   protected Position handlePosition(Position position) {
      Collection<Event> events = analyzePosition(position);
      if (events != null) {
         Context.getNotificationManager()
               .updateEvents(events, position);
      }
      return position;
   }
   protected abstract Collection<Event>
            analyzePosition(Position position);
}
```

After going through this processing chain a `NotificationManager` can propagate the results to other components. For example a SMS gateway (hardware) can be triggered to send a message to the driver of a car.

12.3.1 Traccar's BasePipeline

Another approach to explore Traccar processing is to send a message to a dedicated port and protocol and place a breakpoint. Let's look at our `ProtobufferDeviceDecoder` implementation in the type hierarchy

```
ProtobufferDeviceDecoder extends BaseProtocolDecoder
   BaseProtocolDecoder extends ExtendedObjectDecoder
      ExtendedObjectDecoder implements ChannelUpstreamHandler
```

which is composed of `traccar` classes and implements a `ChannelUpstreamHandler` to connect to the Netty framework.

In order to find a good entry point we'll look at the `BasePipelineFactory` and place a breakpoint in the overridden `.getPipeline()` method. Now you can step through the factory process to see how the decoder is created at the beginning of the pipeline and the handlers are appended.

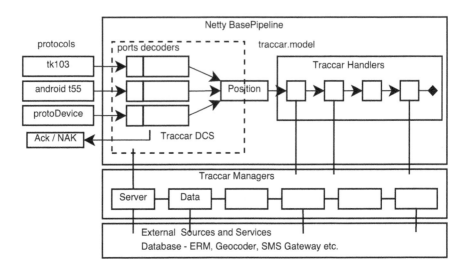

Figure 12.1 The Traccar GTS is running completely in one Netty pipeline and in a single JVM. Traccar handlers are triggered in the pipeline to interact with Traccar managers that coordinate external resources.

If you look inside some Traccar handlers you can see how they interact with different managers via the `Context` configured at program start. For example the `MotionEventHandler` is using `Context.getIdentityManager()` to access the device's latest position.

Once Traccar is running we can use the `jeets-tracker` to send a proto device message. We can place a breakpoint in Netty's `OneToOneDecoder` to see the message coming in from the port. Then the `ExtendedObjectDecoder` receives the `originalMessage` and calls the `ProtobufferDeviceDecoder.decode` method. With `channel.write(ack)` the client receives the Ack and the connection is closed.

At this point you can check the client invocation to find that the connection is closed – while we're still debugging the server. After taking care of the Ack the decoder provides a position entity to be propagated through the rest of the pipeline. As an example you can place another breakpoint in the `GeocoderHandler`, which calls an external service, or the `AlertEventHandler` to see how server events are created. The `MainEventHandler` is the last in line and logs the position.

Again we can look at the type hierarchy of handlers

```
AlertEventHandler extends BaseEventHandler
   BaseEventHandler extends BaseDataHandler
      BaseDataHandler extends OneToOneDecoder
```

to find how Traccar is hooked up to the Netty framework.

Our intention is *not* to go through each one of these managers and handlers. We want to understand how Traccar is going through this complete process inside a single Netty pipeline. And all 'components', like decoders and handlers are extended or implemented from Netty types while the managers take care of the business logic. Our intention *is* to set up a similar processing chain where you can create your own handler components as needed for dedicated applications or use cases.

We have looked at Traccar's event handling process in Section 8.1.2 to find that all events are parsed and created inside its `BasePipelineFactory`. With respect to performance this is hard to beat. With respect to SDC scenarios we would have to implement every new event in the Netty pipeline. As these events can be project specific we would need to add switches, filters etc. inside the streaming process.

Traccar's handler concept analyzes the previously created `Position` *on events derived by the server logic.* Since you can't analyze a motion from a single position at a given time the method takes the previous position into account. Again you should be aware that this previous position is retrieved from existing server data and this data can already have been *smoothed* and deviate from the actual client recording.

By using the device ORM we can submit more than one position and the client can implement some useful logic to compose a track. Then we can simply access each previous position in the track without accessing any external resources.

By creating a persistence unit in conjunction with a protocol we have defined the target format of the complete system. Now we want to go a different way than Traccar by providing a general purpose 'loader' *before* applying any application logic after loading.

We want to pick up the idea of integration frameworks with respect to the Traccar processing chain and perceive every manager as a component which can be added to and removed from the system. This way the software composition becomes much more flexible. Let's see how we can achieve this?

CHAPTER 13

The JeeTS DCS

CONTENTS

13.1   THE CAMEL INTEGRATION FRAMEWORK

Integration technologies in general support the combination of distributed soft- and hardware to one system. For Java there are several integration frameworks available. We will apply the light weight Apache Camel to create a `jeets-dcs` in this and add the Spring framework to create a `jeets-etl` in the next chapter.

The Camel Integration Framework [34]

> "Apache Camel is a small library with minimal dependencies for easy embedding in any Java application. Apache Camel lets you work with the same API regardless which kind of Transport is used - so learn the API once and you can interact with all the components provided out-of-box."

If you are not familiar with Camel it is helpful to browse over its website – no doubt. On the other hand developers confronted with Camel for the first time commonly have problems grasping the frameworks domain, which is somewhere *in between* other software specifications. Integration is the process of *mediating between* different software 'pieces'. Camel is not a middleware itself; it can be used anywhere *between* Java applications, hardware (i.e. ports), middleware and backends. If you look at a block diagram of a system then Camel is represented (or actually hidden) in the connecting lines.

What exactly is ApacheCamel?[1]

> "Apache Camel is an open source Java framework that focuses on making integration easier and more accessible to developers and provides:
> * concrete implementations of enterprise Integration Patterns (EIPs)
> * connectivity to a great variety of transports and APIs
> * easy to use Domain Specific Languages (DSLs)
>   to wire EIPs and transports together "

Camel provides 'a great variety' of different implementations and different domain specific languages, which can be very distracting for a first contact. With Camel you (can) take a different, higher level, perspective to look at complex software systems, to *think* in enterprise integration patterns and to develop in the role of an architect. With plain Java you create Objects to interact in the program flow. With Camel you also code with Java, but now the involved components can be treated analogous to Java objects.

13.2   JEETS CAMEL COMPONENTS

For the readers unfamiliar with Camel the next two chapters can serve as a spectacular introduction to modern software development:

The JEETS Camel Tutorial - get ready for some Camel magic!

The important aspect of integration is to grasp each software component in its business domain language, like device communication. Not in technical terms (TCP, threading etc.) but in how the component works internally.

---

[1]see stackoverflow.com/questions/8845186/

What? Not How?! The JEETS is about creating independent tracking components by combining JEE components with business Code of the domain GPS Tracking. Up to this point we have gained a good insight of the JEE components Protobuffer, Netty and Hibernate, which is the prerequisite to a higher level view on the system.

In Section 12.3 we reflected on the monolithic Traccar application running in a single JVM and processing each message in a single Netty channel pipeline. The challenge to create JEETS server components is the migration of Traccar GTS features and functionality to a distributed environment!

In a *Camel context* messages, i.e. *Camel exchanges*, are processed in various steps when traveling through a *Camel route* with well defined start- and *endpoints*. A route is nothing really new to us. In low level TCP/IP, Section 2.2), we speak of sockets and threads. After a client starts a thread to send the message the server also starts a thread to handle this message and send a response. Upon successful message transmission all threads are closed. In the context of Netty we speak of channels and pipelines – the same thing, but a higher level.

By routing the message in Camel we can implement the same processing steps, same logic, same filters that make up the Traccar GTS. We could even transfer a Traccar *Manager code (see page 124) to a single Java component. We could filter notifications and send them to a driver notification module – or whatever!

13.3   IMPLEMENTING WITH CAMEL

We are focused on the JEETS and we'll introduce the Camel terminology as we implement the JEETS DCS step by step. It is up to you to clarify these terms and your missing pieces online. We will go through everything we have covered for the client software – again.

Before you complain "Not again" – relax. This time we will apply even *less* code to achieve the same functionality *faster*. The DCS will be implemented with plain Camel and completely in Java. For an ETL module we'll look at the Spring framework and how it can *reduce the code even more* in conjunction with Camel.

13.3.1   Components

Components are the main topic of this book (see title) and GPS tracking is our problem domain. Since you decided to read this, you already have established a picture of what components are and Section 2.3 provides our contract to work with. From the Camel view components are treated as black boxes with an input and output of some kind of data in some kind of format – not restricted to Java nor by hardware.

Camel is widely accepted and provides new Camel-ized component implementations continuously. On the Camel website you can find a few hundred out-of-the-box components [35] ! One of these components is the Netty framework [36], a proven component in the Java community and a component we have studied in some detail.

Naturally the developer has to know what happens inside a black box in order to integrate it ('hook it up') with other black boxes, i.e. components. From our Netty studies we know that Netty consumes messages from a port and decodes them for further processing in the system. That's a good start to get familiar with Camel.

13.3.2 Camel Setup

Camel wouldn't be a lightweight framework with a small footprint if all of the out of the box components were part of one project. Therefore you always need the Camel core and the component you want to use. Please have a look at the JEETS DCS component's pom file:

```
<dependency>
    <groupId>org.apache.camel</groupId>
    <artifactId>camel-core</artifactId>
</dependency>
<dependency>
    <groupId>org.apache.camel</groupId>
    <artifactId>camel-netty4</artifactId>
</dependency>
```

As you can see the dependency configuration is not more complex than setting up the tracker; it's actually less complex. The DCS pom file includes a `<dependencyManagement>` for Camel in general[2] to manage Camel components with predefined versions. Consequently there is one single place to update Camel with all of its components. For example we don't have to version Netty at all and the complete component is automatically loaded by Maven. Also note that the Netty framework *is not included* in the `jeets-protocols` and is automatically added by Camel.

When creating your pom file you should be aware of implicit artifacts. By importing the `jeets-protocols` project we decided earlier to include the `jeets-pu` project with JEE components Hibernate and Protobuffers. In case of a version conflict with Camel it makes sense or may become necessary to alter the order of dependencies to prioritize one version.

13.3.3 Component Endpoints

Developer's Note:
To support the development of the first JEETS server component we have created the JEETS tracker to send device messages. In development time we

---

[2]see Maven BOM, Bill of Materials

can transmit the message to our Traccar Installation, switch the server port and send it to the new JEETS DCS for validation.

Next is *the* fundamental term of integration: The component endpoint.

You can picture the server port as the (hardware) endpoint as defined on page 13, in this case the server *input* point, where server processing starts. We have picked the Netty framework to hook up to the port and take care of TCP communication, threading, blocking etc. From Camel's perspective Netty is yet another JEE component. You can define an endpoint in Camel syntax with

```
from("netty4:tcp://localhost:5200 .."
```

We can read this like "Camel please use Netty, version 4 in TCP mode to hook up to the local host and listen to port 5200.". You can now run the `org.jeets.dcs.Main.main` method and have a look at the startup sequence in the logging output:

```
INFO org.apache.camel.component
             .netty4.SingleTCPNettyServerBootstrapFactory
    - ServerBootstrap binding to localhost:5200
INFO org.apache.camel.component.netty4.NettyConsumer
    - Netty consumer bound to: localhost:5200
```

The package structure `camel.component.netty4` indicates that Netty4 is a component accessible via Camel. `ServerBootstrap` sounds familiar from page 21 among others. Camel and Netty take care to set up a port listener.

No programing required!

Now we don't have to deal with the incoming message on the TCP level any more. Implicitly we connect the hardware endpoint to the Netty startpoint (consumer) and then want to pick up the decoded protobuffer message at the Netty endpoint (producer). This is another important distinction we need to keep in mind: endpoints can be subdivided into input `from`- or output `to` points. A Camel route is defined by a processing sequence between `from` and `to` endpoints. To underline this more than subtle difference we will speak of starting-, ending- or endpoint for either one.

Internally a starting point is always a different Camel endpoint implementation than an ending point. A starting point is always handled internally by a dedicated Camel `<Component>Consumer`. What you need to understand is that the Netty starting point is completely hidden inside the component. By defining the endpoint `from("netty4: ...)` we are actually hooking up to Netty's output, i.e. ending point, which is implemented in a `<Component>Producer`. As a Camel user you don't necessarily have to deal with these internals, but without the distinction you can get lost easily.

For a better understanding we don't want to accept Netty as a black box. We will dig a little deeper to get familiar with Camel concepts. As we have become familiar with Netty we have the knowledge that Netty includes a protobuffer decoder (see page 91) that we want to apply to our device message. If this happens inside Netty how can we interfere with the component?

Camel per se can not customize a component for your needs and provides a variety of domain-specific languages (DSLs) for individual configurations [36]. Therefore we are using the protobuffer decoder inside Netty `io.netty.handler.codec.protobuf.ProtobufDecoder`. We are familiar with this solution on the client side and it allows us to migrate the Traccar protocol implementations to the JeeTS DCS component, if needed.

With Camel you always have many choices and another approach for a DCS implementation would be to use Camels Protobuffer component [37]. This is typical source of confusion in Camel coding. For clarification the protobuffer component is defined for the test environment only and you can run the `ProtobufNettyCamelTest` to see how it works.

You could also choose to use the Netty component without built in protobuffer decoder and use the protobuffer component to un/marshall protobuffer binaries. This would be the obvious implementation for a `jeets-dcs`, but as mentioned we want to use the Traccar-Netty pattern in order to add Traccar protocol implementations.

13.3.3.1 Component Configuration

The complete syntax of the configured entry point looks like this

```
from("netty4:tcp://localhost:5200
                ?serverInitializerFactory=#device&sync=true"
```

and works like URL parameters with `key=value` combinations. The parameter `sync=true` defines the protocol as bidirectional and prepares the response with our predefined `Ack` message. The

```
org.apache.camel.component.netty4.ServerInitializerFactory
```

also sounds familiar and is a Netty customized Camel class or a Camelized Netty class ;) It works exactly like the actual Netty decoder with the slight difference of the additional `NettyConsumer` to consume the incoming message. The `NettyConsumer` represents the consuming endpoint inside the Netty component. We have done this before and simply add our decoder pipeline to create a `DeviceProtoExtractor`:

```
chPipeline.addLast(new ProtobufVarint32FrameDecoder());
chPipeline.addLast(
        new ProtobufDecoder(Traccar.Device.getDefaultInstance()));
chPipeline.addLast(new ProtobufVarint32LengthFieldPrepender());
chPipeline.addLast(new ProtobufEncoder());
chPipeline.addLast(new ServerChannelHandler(nettyConsumer));
```

which includes our protocol, i.e. protobuffer type `Traccar.Device` and a `nettyConsumer` instance to connect to the component's internals. By applying this pattern it should become clear how to migrate a Traccar protocol implementation to the DCS.

For this book we are working with an initial `Traccar.Device` message to demonstrate the usage of protobuffers in conjunction with a system model, the PU. For your purposes you can use the Traccar model to create additional messages for any aspect of the model. With this constellation you can define a port with any proto message or proto file as a complete message catalog in no time. In Section 2.1.3 we have looked at Garmin's FMI Message Catalog defined on 70 pages. This catalog could be modeled with a number of related `.proto` files and by defining a top level message you could use one single port to submit any Garmin FMI message. Or you could define one port for each message, etc.

13.3.4 Building Routes

The syntax `from("netty4:...")` defines the Netty output point. At the same time this is our Input point at the start of the route to receive the incoming messages. What's next? How can we process the message? In Camel jargon the message is traveling through a *Camel route* and can be modified in any number of steps.

For the `jeets-dcs` Camel route you can locate the explicit `DcsRoute` class. For Camel it makes no difference where you configure your `RouteBuilder` and sometimes you may prefer an anonymous class for better readability. Let's look at the DcsRoute configuration:

```
public class DcsRoute extends RouteBuilder {
   public void configure() throws Exception {
      from("netty4:tcp://localhost:" + PORT
         + "?serverInitializerFactory=#device&sync=true")

         // TODO: add processing steps

      .to("XXXX:jeets-dcs")
   }
}
```

This class defines the data flow, i.e. Camel route for the `jeets-dcs` component and `RouteBuilder` is always the place to look at and get a quick impression of what a component essentially does. In the listing you can see the configured Netty endpoint and `from` this point we are expecting an extracted `Traccar.Device` message from the implied `DeviceProtoExtractor`. The `"XXXX:jeets-dcs"` is a placeholder for the time being. We will add this `to`-point a little later in Section 13.3.9.

13.3.5 Message Translator

As a message is traveling along a Camel route from one component to another it can or must change its format to comply with the next component! Format change is an implicit requirement for any routing, i.e. integration.

The EIP message translator definition [32] modified for our context states that a `Traccar.Device` message carries information that we want to 'translate' to a system `Device` entity. Two different Java types with *almost* identical context as we have analyzed in Section 7.4 Data Tranformations. Camel provides many patterns to achieve this with subtle differences:

> Data Format
> If we were not using Netty to decode the protobuffer payload we could receive the binaries from Netty and unmarshall them.
>
> Message Translator
> This mechanism could be used to transform the protobuffer message content to a **new** system entity.
>
> Type Converter
> Use this to change the way of handling the message content, i.e. from a String to a Stream.

Now we can start adding the processing steps for device communication. In a DCS (and later ETL) context we want to transform a `Traccar.Device` message to a `Device` entity and we have already written the transformation code in Section 7.4 ! We'll choose the type conversion and apply Camel's type converter mechanism [38] in three steps:

> First we create a simple Java class `DeviceProtoToEntityTransformer`
>
> where we can access the `org.jeets.protocol.util.Transformer`; then
>
> we add the camel `@Converter` annotation to the method:

```
@Converter
public final class DeviceProtoToEntityTransformer {

    @Converter
    public static Device toDevice
        (Traccar.Device deviceProto, Exchange exchange)
                                    throws Exception {
      return org.jeets.protocol.util.Transformer
                    .protoToEntityDevice(deviceProto);
    }
```

The only Camel constructs are the `@Converter` and the `Exchange` in the method signature, where the first parameter is the from-type and -value while the return is the to-type and -value.

After we have created the `DeviceProtoToEntityTransformer` we need to tell Camel where to find it. This is described on Camel's `type-converter.html` page under the heading "Discovering type converters". The `TypeConverter` file with the qualified classname is placed in the `META-INF/services/org/apache/camel/` folder. And finally we can add the transformer to the `DcsRoute` with[3]:

```
.convertBodyTo(Device.class)
```

After this transformation step we can pick up the entity and forward it to the system. On the other hand we have configured the Netty endpoint to operate in request-response (`sync=true`) mode for an acknowledged transmission.

At this point we are going for a different processing of entities than Traccar. The main concern for performance here is to return the `Ack` directly after extraction and transformation to release the connection. The subsequent processing of the message should be decoupled from network operations to avoid network and client blocking and waiting for external sources.

13.3.6   Processor

The processor interface belongs to the core of the Camel concept. After defining the `from`- and `to` endpoints of a route you usually want to do something with the exchange data and a processor implementation is something like a `between` point. In general you can assume that any 'mechanism' provided by Camel, like a type converter, is implemented in some kind of processor.

Developer hint:
When you want to 'look into' a route you can always insert a temporary processor in your route to do some debugging to get a clear picture:

```
.process(new Processor() {
   public void process(Exchange exchange) throws Exception {
      // debug breakpoint, logging, println etc.
   }
})
```

By implementing the processor interface you can easily code your individual `.process` method with plain Java and the Camel framework provides access to an `Exchange`. On an entry level you don't have to deal with the exchange directly: For the transformer we didn't look at the implementation details at all and by simply adding a signature and it works!
The `jeets-dcs` implements a simple processor to acknowledge messages:

```
public class AckResponder implements Processor {
   @Override
   public void process(Exchange exchange) throws Exception {
```

---

[3]With a message translator it would be: `.transform(body())`

```
        // implicit Type Conversions
        Device devEntity = (Device)
               exchange.getIn().getBody(Device.class);
        // Traccar.Device devProto
        // = exchange.getIn().getBody(Traccar.Device.class);
        // = (Traccar.Device) exchange.getIn().getBody();

        //      TODO: validate transformation:
        //      if (valid) return ACK, else return NAK

        Traccar.Acknowledge.Builder ackBuilder
               = Traccar.Acknowledge.newBuilder();
        ackBuilder.setDeviceid(devEntity.getUniqueid());
        exchange.getOut()
        .setBody(ackBuilder.build(), Traccar.Acknowledge.class);
    }
}
```

All you have to do is grab Camel's `Exchange` to pick up the `Traccar.Device` and / or the `Device` entity, do some processing to create the `Traccar.Acknowledge` message and set it on the exchange.

The commented lines demonstrate a useful 'side effect' of the build in type converter. If you are interested you can access the device entity and device proto and transform them by casting one to another. It is a pure design decision to move the transformation to the `AckResponder` or use an explicit statement in the route as we have done earlier. In Camel you should never look for *the one* correct implementation. It provides sufficient freedom to develop your personal coding style.

Message validation is only indicated in the listing and can be anything from a simple plausibility test of longitude and latitude to analyzing a complete sequence of positions. Any validation at this point is part of the network transfer and should return an `Ack` or `NAK` (see page 16) quickly.

This quick validation is also the place to reject messages. Currently all major enterprises in the automotive industry are creating their own SDC backends for selected OEM (car vendors). There is no official or governmental framework to collect signals from every car on every road. The problem is that every dedicated SDC scenario has to implement some kind of fast authorization to reject unwanted GPS messages and to accept cars equipped for this scenario.

13.3.7 Message Exchange

It wouldn't make sense to set up a processing route without data flowing through it. Camel was designed to handle any message or more generally any Java type in a route. This design is based on a number of message exchange patterns like request-reply or events that only travel in one direction.

The Messaging Pattern [33]

> "An enterprise has two separate applications that are communicating via messaging, using a message channel that connects them. How can two applications connected by a message channel exchange a piece of information? Package the information into a message, a data record that the messaging system can transmit through a message channel. Thus any data that is to be transmitted via a messaging system must be converted into one or more messages that can be sent through messaging channels."

Our `Traccar.Device` message is implicitly wrapped in a Camel `Exchange` type which is used inside the Camel core. Camel provides the `Exchange` via the `.process` method and with `exchange.getIn()` you can access the device entity coming `In`.

The `.process` method does not define a return value as a Java method parameter object is called by value. You can implicitly set the return value with `exchange.getOut().setBody(..)` to allow the Camel framework to propagate this message to the next endpoint. Here is another catch: By invoking `exchange.getOut()` Camel implicitly sets some out message and to avoid this you should first check if `exchange.hasOut()` [39].

13.3.8   Message Direction

We have studied how the Traccar GTS processes each message in one Netty pipeline. The Ack message is written back to the client by `channel.write(ack)` while the handlers are invoked in the Netty upstream. To decouple the network communication from the system we want to acknowledge the reception of a message immediately and synchronously propagate the message to the system.

Haven't we just created an `AckResponder` to achieve this?

Yes and no.

Yes, we have created the `Ack` message and it is 'in place'
to be returned to the client via its TCP channel.

No, it is not returned by the `AckResponder`.

Without further measures the `Ack` message will simply reside in the exchanges out 'pocket', but keep on traveling up the route. If we would add more endpoints and processors to the route, like persisting the message in a database, we would have to wait for a successful transaction before the message would travel back down the route to the client. And even worse the route might be linked to one or more other routes and our `Ack` message would also traverse them up and down.

In order to acknowledge the message right after network validation we simply change our endpoint of the route from `.to` to `.inOnly`.

13.3.9   Camel Endpoints

Where does a route end? What happens at the end of a route?

By definition a DCS should supply a system entity, i.e. Device as output and we can represent this output with a .to (or .inOnly etc.) endpoint. This endpoint is defined by the name of "jeets-dcs". We have almost reached the end of the DcsRoute and only need to fill in the placeholder .inOnly("XXXX:jeets-dcs") for the type of endpoint. The general impediment during the development of components is that we are not accessing this output by a 'next' endpoint defined in some route.

The DcsRoute defines the complete device communication and provides the output to the system at its ending point. The Camel way is to create another route in another component that begins with the DCS ending point to decouple the components. We will create a subsequent ETL component to demonstrate this in the next chapter.

The creators of Camel, the Camel riders, have provided ready to use endpoint implementations for seamless integration via (yet unknown) routes [62]. Each of these endpoints only accept Camel exchange messages and operate in-memory, i.e. it has no persistence backup in case of a system crash. The simplest of four Camel endpoints[4] is called direct and we can now finalize the DcsRoute:

```
from("netty4:tcp://localhost:" + PORT
    + "?serverInitializerFactory=#device&sync=true")
.convertBodyTo(Device.class)
.process(new AckResponder())
.inOnly("direct:jeets-dcs");
```

Four lines of code for a device communication server! [5]

13.3.10   Multi Threading

Camel's direct endpoint is designed to accept the data from the route and direct-ly invokes a consumer in a next step or route. Please recall that the Device protobuffer is not just a simple message string. It is an object relational model with unknown size.

Let's imagine we receive a device ORM with many positions and the validation goes through all of them to check chronological order and to remove invalid GPS messages – which takes some time. Right after this message an important warning event arrives and has to be propagated to a car traveling at 50 miles per hour. The traffic situation requires the car to slow down as soon as possible. Now our system is still processing the large ORM, while the small warning message is still waiting for resources.

As a programmer you think of multi threading and the solution is to kick

---

[4]direct, seda, direct-vm, vm
[5]Or three lines by implying the transformation in the AckResponder

off the large process and immediately accept the smaller warning message. Multi threading is a programming pattern and, guess what, Camel provides multi threading out of the box. The `direct` endpoint uses no threading and invokes every consumer one message after the other, while the...

SEDA Component [40]

> "...provides asynchronous SEDA behavior, so that messages are exchanged on a BlockingQueue and consumers are invoked in a separate thread from the producer.
>
> Note that queues are only visible within a single CamelContext. If you want to communicate across CamelContext instances (for example, communicating between Web applications), see the VM component.

Instead of waiting for messages to be sequentially consumed from the end of our route we want serve each consumer (instance!) in a separate thread. Just like other Camel components the SEDA component provides configuration options and now we can fine tune our route for asynchronous concurrent consumers. By specifying `concurrentConsumers` we can improve the performance and with `waitForTaskToComplete=Never` SEDA will simply forget the message call and perform the next one: "fire and forget".
Now the configured multi threaded ending point of the DCS looks like this:

```
.inOnly("seda:jeets-dcs" +
    "?concurrentConsumers=4&waitForTaskToComplete=Never");
```

The simplicity of Camel configurations can easily obscure the complexity of its backing implementations. By setting the number of `concurrentConsumers` you should be aware of the weight of creating threads in a system and determine a reasonable number for the production system.

13.3.11 Camel Context

Now that we created our `DcsRoute` and can't wait to see it in action. In order to apply Camel you need to start a `CamelContext` to let Camel start up components and endpoints and validate the routing rules before the route is ready to receive data.
Before starting the Camel context we need to make the involved classes visible to Camel by putting them to a registry. In our Netty configuration we have used `serverInitializerFactory=#device` to configure the device protocol. The class `DeviceProtoExtractor` name describes that the protobuffer `Traccar.Device` message is extracted from the binary network format, is defined in the URL configuration and must be registered by the name `device` in a `CamelContext`. The easiest way to achieve this for the DCS component is to create the usual starting point of a Standard Java application, the `main` method:

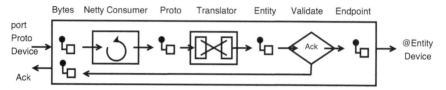

Figure 13.1 Enterprise integration patterns applied by Camel for the jeets-dcs. message data and formats are propagated inside Camel exchanges.

```
public static void main(String args[]) throws Exception {
    SimpleRegistry registry = new SimpleRegistry();
    registry.put("device", new DeviceProtoExtractor(null));
    CamelContext context = new DefaultCamelContext(registry);
    context.addRoutes(new DcsRoute() );
    context.start();
}
```

Now you can launch the DCS and then send a device message with the jeets-tracker from command line. Note that the DCS is not configured as a standalone jar with dependencies. Why would you start a DCS without a consumer? [6] The DCS is a tiny component, but it also has a clearly defined domain and by creating a separate Maven project it can be used in many contexts as we will see soon.

Yet you should already have a feeling for Camel's power to create software. In the last few pages we have implemented Netty with protobuffers in ridiculously few lines and with high maintainability. In large SDC environments you have many (many many) ports and their URL configuration can be achieved with string concatenation – with Java beginner skills. The architect provides a single protocol as a template or pattern for the complete system.

13.4 DCS CONFIGURATION

After creating a component it makes sense to go through its configuration options, since this should be the only concern for (external) component users. We have configured Netty and SEDA with

```
from("netty4:tcp://localhost:1234"
    + "?serverInitializerFactory=#device&sync=true")

.inOnly("direct:jeets-dcs" +
        "?concurrentConsumers=4&waitForTaskToComplete=Never");
```

---

[6]You could use the VM component...

which we can translate to

```
( host=localhost )
  port=1234
  protocol=device          ( with Transformer class )
  concurrentConsumers=4
```

and actually we can remove the host and hardwire it to localhost. For a Java developer it shouldn't be a problem to integrate three configuration parameters in his code. Then the system developers are not confronted with Netty and protobuffer when they want to add a DCS for their `proto`-col and port! Note that the protocol `device` is only a lookup key for the registry and therefore the developer is even free to register a different transformer class in his environment.

Now you can start the DCSs `Main.main` method and, if you have precompiled the JEETS repository, run the tracker's main method to send device messages. You should see an ACK returned for every message. Or you can start the tracker first to collect some messages to be sent and then start the DCS to receive them.

The JeeTS ETL

CONTENTS

It is a DC server's domain to consume network messages and to transform them to system entities. In Section 2.2.2 we have looked at OpenGTS where the DC servers are packaged in independent jar files that can be launched on the command line. This is useful for the production operator to leverage network traffic. Each process is visible on the OS level with unix commands like `top` and `ps`. These small jar applications store the decoded messages in the database. Many GTS features, like admin- and map frontend and reporting engine are coded against the database without awareness for device communication.

In enterprise computing this approach is usually implemented in an Extract-Transform-Load, ETL component[1]. Inside the DCS we already have *extracted* (E) the protobuffer binaries and *transformed* (T) them to entities,

---

[1]en.wikipedia.org/wiki/Extract,Transform,Load

so now we have to add the *'loading'* (L). Although an ETL module can include a DCS (for E and T) it is important to distinguish between them.

In SDC scenarios databases are very unpopular due to comparably slow hardware access for very high data rates streaming in. For small environments we could create in memory databases with our PU. This is demonstrated in the test environment and you can determine performance and should consider the time span to create a database schema (ERM) in relation to the total database lifetime. Why store every bit of information, if you want to monitor the *current* traffic? Forget the past.

On the other hand we want to apply our new components to our production system Traccar. We will use this constellation to create an ETL module by importing the DCS. This way we can find out how to combine Camel routes in modules and provide a nice scenario to introduce the Spring framework. By persisting our JEETS messages to the Traccar database we can collect messages from the `jeets-tracker` client with the `jeets-dcs` server, persist them with the `jeets-etl` and continue using the administrative frontend to register trackers, to visualize their routes on a map or list them in reports.

14.1   THE SPRING FRAMEWORK

The two main reasons to apply the Camel and Spring frameworks are their threading and Decoupling capabilities under the hood. Both techniques are applied to avoid delays or blocking, i.e. wasting performance in the regular program flow. Developers describe these low level technical terms with 'Context and Dependency Injection' (CDI and DI). Since the reduction of coupling is vital for complex systems the community strove for a lightweight Inversion of Control (IoC) container for simple POJOs and beans. While Camel is decoupling components, Spring was initially made to decouple objects, i.e. POJOs or beans. The Spring framework is an dependency Injection framework to manage the life-cycle of Java components and/or beans.

The Spring Framework [63]

> "Core support for dependency injection, transaction management, web applications, data access, messaging, testing and more.
>
> The Spring Framework provides a comprehensive programming and configuration model for modern Java-based enterprise applications - on any kind of deployment platform. A key element of Spring is infrastructural support at the application level: Spring focuses on the "plumbing" of enterprise applications so that teams can focus on application-level business logic, without unnecessary ties to specific deployment environments."

When annotations were added to the JEE specification v6 as an alternative to xml configurations Java EE began to break out of the application server limitations, which were designed to provide a complete JEE implementation. For smaller processes this turned out to be a massive overhead. With the Spring framework you can define independent Java objects and services and wire them together.

If you look at the Spring framework Quickstart [64] you will find an introductory example of a `MessageService` and a `MessagePrinter`. If you would create a `MessagePrinter` as a member of the a `MessageService` instead, the service would be allocated for the complete time of printing. Imagine you would call the service to print a very long report and call the service again while the report is still being printed. Since this allocation is a waste of performance and time, the idea is to decouple the objects and create new instances when ever needed. In this case Spring will create and *inject* a `new MessagePrinter()` to print another report parallel to the first.

In the JEETS context we could think of one Java bean to gather analyzed GPS routes by implicitly calling an `.analyzeRoute( deviceORM )` method. Now different cars from different OEMs can implement different rules to transmit one device ORM with only a few or many positions. Naturally the route analysis would be more or less complex and require different durations.

Creating the `jeets-dcs` was used to introduce Camel terminology and concept. The application was deliberately created in plain Java. With Spring we can apply xml configurations and Spring annotations to our Camel modules. In a simple picture Spring provides the runtime environment to configure and run Camel with auto configuration and to provide additional 'glue' to support the components involved. Have a look at `camel.apache.org/spring.html` and we'll take it step by step.

14.1.1 Camel Spring Setup

We have already setup Camel for the `jeets-dcs` and now we'll import the complete project into the `jeets-etl` project. With Maven we can add dependent JEETS components like the persistence unit with drivers, protocols with protobuffers and the DCS with Camel core and Camel Netty. In order to create a 'loader' module we will add the Spring framework and later we will add a JPA implementation to actually access the database.

```
<dependency>
    <groupId>org.apache.camel</groupId>
    <artifactId>camel-spring</artifactId>
</dependency>
```

Keep in mind that you should define a single place in your pom hierarchy to alter the `${camel.version}` for all components used by Camel.

14.1.2   Startup and Shutdown

For the `jeets-dcs` we have defined a `Main` class in Section 13.3.11 to run the application – only for development and testing. When importing the DCS project we will simply ignore this class to avoid the creation of a registry, a Camel context and the Camel route.

For the `jeets-etl` we will also add a `Main` class which is only created to materialize the `org.apache.camel.spring.Main` class:

```
public class Main extends org.apache.camel.spring.Main {
    public static void main(String... args) throws Exception {
        new Main().run(args);
    }
}
```

and that's all! What? The javadocs provides the brief description "A command line tool for booting up a Camel context using an optional Spring application context."

This time we are going the opposite way of creating the DCS. First we create the `Main.main` class and method as the starting point from the command line and then we will fill in the pieces for the ETL module. For a first impression you can go to the ETL directory and enter the following command line with or without testing:

```
repo.jeets/jeets-etl> mvn camel:run (-Dmaven.test.skip=true)
```

Maven is configured with the `camel-maven-plugin`[2] to compile (, test) and package the ETL module. Then Spring will auto discover the configuration (that we will go through step by step) and launch the module.

```
Using org.apache.camel.spring.Main to initiate a CamelContext
starting Camel ...
...ClassPathXmlApplicationContext prepareRefresh
...XmlBeanDefinitionReader loadBeanDefinitions
Loading XML bean definitions from file
    [...\jeets-etl\target\classes\META-INF\spring\camel-context.xml]
[org.apache.camel.spring.Main.main()]
  - Apache Camel 2.19.0 (CamelContext: camel) is starting
```

In the output we can see that the configuration resides in a `camel-context.xml` file. You can achieve the same startup listing in your IDE by launching the `Main.main` method. But then you can not witness the graceful shutdown when you stop the `mvn camel:run` with `<Ctrl><C>` on command line:

```
Received hang up - stopping the main instance.
org.apache.camel.main.MainSupport exiting code: 0
```

---

[2]You can also use the faster `exec-maven-plugin` which specifies the Main class to run 'mvn compile exec:java' without packaging the project.

```
Apache Camel 2.19.0 (CamelContext: camel) is shutting down
Starting to graceful shutdown 2 routes (timeout 300 seconds)
ServerBootstrap unbinding from localhost:5200
Netty consumer unbound from: localhost:5200
Route: route2 shutdown complete from: tcp://localhost:5200
Route: route1 shutdown complete from: seda://jeets-dcs
Graceful shutdown of 2 routes completed in 0 seconds
CamelContext: camel has been shutdown,
               triggering shutdown of the JVM.
```

We can see what Spring's `Main.main` is doing for us – without writing an explicit method body! The production manager can start and stop the ETL while Spring and Camel are checking for 'inflight' messages, waiting for them to get processed and then shutting down without losing a message. Very nice.

14.1.3   Camel Context

If we are not using the DCS's `Main.main` method to startup then how does Spring know how to configure the DCS inside the ETL? The startup listing provides the answer [41]:

```
Loading XML bean definitions from file
   [...\jeets-etl\target\classes\META-INF\spring\camel-context.xml]
```

Generally you can also define the context in the `main` method with

```
ClassPathXmlApplicationContext applicationContext =
   new ClassPathXmlApplicationContext("camel-context.xml");
```

which explicitly loads the Camel context into a Spring application context. In this context the name `camel-context.xml` was chosen although Spring allows different kinds of xml names. Let's look at this file and find out how to configure DCS and ETL. To make the file Spring and Camel aware, their schemes are added:

```
http://www.springframework.org/schema/beans
http://camel.apache.org/schema/spring
```

with their xsd locations. The Camel context is defined with

```
<camelContext id="camel" trace="true"
         xmlns="http://camel.apache.org/schema/spring">
   <routeBuilder ref="EtlRoute"/>
   <routeBuilder ref="DcsRoute"/>
</camelContext>
```

In a plain Spring context the two routes refer to beans that must be registered with fully qualified names:

```
<bean id="EtlRoute" class="org.jeets.etl.EtlRoute" />
<bean id="DcsRoute" class="org.jeets.dcs.DcsRoute" />
```

and we can do the same with the Netty Extractor that needs to be registered for the URL configuration `serverInitializerFactory=#device`:

```
<!--
<bean id="device" class="org.jeets.etl.steps.*ProtoExtractor">  -->
<bean id="device" class="org.jeets.dcs.steps.DeviceProtoExtractor">
   <constructor-arg name="consumer">
      <null />
   </constructor-arg>
</bean>
```

The `DeviceProtoExtractor` only defines the `Traccar.Device` protocol and the interesting part is that we can choose to use the (default) protocol of the DCS or we can add a different *extractor* inside the ETL module to apply a different protobuffer protocol. Currently the `etl.*Extractor` is commented and you can activate it for your .protocol file. Since the extractor does not supply a default constructor you have to define the argument name matching the name in the Java code and set the argument to null like we did in the DCS startup.

14.1.4   Combining Camel Routes

In the xml configuration we have simply listed the `DcsRoute` from the DCS and added an `EtlRoute`. Open the latter to find the `from("seda:jeets-dcs")` as the ETL routes starting point, which is the DCS ending point. Also note that the

```
class EtlRoute extends SpringRouteBuilder
```

and have another look at the startup sequence:

```
SpringCamelContext
   - Route: route1 started and consuming from: seda://jeets-dcs
...SingleTCPNettyServerBootstrapFactory
   - ServerBootstrap binding to localhost:5200
...NettyConsumer
   - Netty consumer bound to: localhost:5200
SpringCamelContext
   - Route: route2 started and consuming from: tcp://localhost:5200
   - Total 2 routes, of which 2 are started.
```

That's all we need to direct the DCS output to the ETL input!
Now we can create the ETL route.

14.2   CAMEL JPA COMPONENT

The ETL adds database persistence to the DCS and the `jeets-etl` project houses the configuration for the JEE JPA API and its implementation. JPA was specified as a layer to abstract the object/relational mapping implementations. In order to verify that the `jeets-pu-traccar-jpa` adheres to the JPA specification we will use Camel's `camel-jpa` [42] component to interact with the Traccar database.

14.2.1   Configuration and Implementation

While the API is sufficient to start coding we need an implementation to run the ETL. We will choose the OpenJPA implementation, but if you would like to work with `jeets-pu-traccar-hibernate` and Hibernate `.hbm.xml` files you might as well use Camel's `camel-hibernate` [43] component.

```
<dependency>
    <groupId>org.apache.camel</groupId>
    <artifactId>camel-jpa</artifactId>
</dependency>
<dependency>
    <groupId>org.apache.openjpa</groupId>
    <artifactId>openjpa</artifactId>
    <version>${openjpa-version}</version>
</dependency>
```

The configuration resides in the `camel-context.xml` file to declare relevant helper beans for Spring. They are exposed via registry and can be dependency injected as components, endpoints etc. The configuration also takes care of database transactions. As stated on the Camel Spring page Camel uses Spring transactions as the default transaction management in components like JMS and JPA. This is reflected by their full qualified names:

```
<bean id="jpa" class="org.apache.camel.component.jpa.JpaComponent">
    :
<bean id="entityManagerFactory"
    class="org.springframework.orm.jpa.LocalEntityManagerFactoryBean">
    <property name="persistenceUnitName" value="jeets-pu-traccar-jpa"/>
    :                                            --------------------
<bean id="transactionManager"
    class="org.springframework.orm.jpa.JpaTransactionManager">
    :
```

It is not important *where* you define the PU name but you should be aware that every component of the JEETS is coded against the system model defined in the PU. Now we can finally define the ending point of the ETL component with:

```
.to("jpa");
```

By optionally adding the fully qualified entity Name explicitly to the URL

```
.to("jpa:org.jeets.model.traccar.jpa.Device");
```

we can add type Safety explicitly for our proto device inside the ETL component.

14.2.2   Camel JPA Configuration

Now that we have an endpoint for persistence, the ETL should be ready to use. But again we are confronted with the intricacies of a Camel component. When programming against the JPA you must be aware of the differences between the `EntityManager`'s `persist` and `merge` methods. Behind the scenes the difference lies in the SQL statement to UPDATE or INSERT data.

The `persist` method adds a new entity instance to the persistence context and has the `void` return type. The problem is that you can raise an exception if you try to persist a *detached*, i.e. *unmanaged* instance with something like `entityManager.persist(entity)`.

The `merge` method looks for an entity via ID, retrieves it from the persistence context or creates a new instance loaded from the database, copies fields from the passed object to this instance and returns the updated instance. The returned entity is a different object and the passed-in object should be ignored – or better discarded.

In the context of the ETL in conjunction with Traccar we would like to first check, if a device is registered under the `uniqueId` and only persist the message for existing devices. This way we can still use the Traccar administration to register devices and then receive their messages via ETL.

Applying SQL statements or an `EntityManager` on a database is more than just a technical detail. Database models (ERM) are intentionally created with a number of constraints to control data modification (DML). Therefore we will choose the `persist` method to gain control over the incoming data. Checkout the Camel JPA page to find the configuration option `usePersist=true` to append to the endpoint

```
.to("jpa:org.jeets.model.traccar.jpa.Device?usePersist=true")
```

where the body of Camel's In message is assumed to be an `@Entity Device` bean.

14.2.3   Message Translator and Data Access

The actual persisting in JPA is always coded against an `EntityManager` and additional handling code is defined by annotations in the persistence unit. We have used Hibernate Tools to create a `jeets-pu-traccar-jpa` in Section 6.3 and have added some code to the `@Entitys` like `CascadeType` earlier. To interact with the database via the Camel JPA component in the Java code we need to get hold of an `EntityManager` to work with.

Please have a look at the `camel.apache.org/etl-example.html` which is close to the solution we are looking for. In the `CustomerTransformer` listing you can locate the entity- and transaction management:

```
EntityManager entityManager = exchange.getIn()
    .getHeader(JpaConstants.ENTITYMANAGER, EntityManager.class);
TransactionTemplate transactionTemplate = exchange.getContext()
    .getRegistry().lookupByNameAndType
        ("transactionTemplate", TransactionTemplate.class);
```

The `camel-example-etl` transforms a `*.xml` file to a `CustomerEntity` and the `CustomerTransformer` implies database transaction. In our case we have already transformed the device proto to a device entity in the DCS and actually wouldn't need an explicit transformer any more. We will use one anyway to learn how to access Camel component internals by applying a little trick.

The combined DCS and ETL routes start with the Netty component and end at the JPA component. If we would place a `Processor` right before the JPA endpoint then the above code to retrieve entity- and transaction beans would return `null`. We can conclude that the intermediate `Processor` *is not aware* of the components at the route endpoints.

So we'll look at Camel's `etl`, `jpa` and `hibernate` pages closer to find a little hint: "If the body does not contain an entity bean, use a message Translator in front of the endpoint to perform the necessary conversion first." The lesson to learn is that the JPA component automatically triggers a type converter, if the incoming class is not a device entity. Now the type converter *is aware* of the persistence context!

We want to (miss-)use the type conversion to transform the device entity from the network to a potentially existing device entity in the database. We implicitly have to lookup the `uniqueId` and return a *managed* device entity.

Transforming a device to a device doesn't make sense and gives Camel no chance to detect a valid use case in the code. So we will create an 'artificial' class `NetworkDevice` to convert from. The additional class is only a wrapper class to hold the incoming device as a member and the transformation simply calls the method `getDevice()`. The conversion from `Device` to `NetworkDevice` is prepended in a `Processor`.

```
Device device = (Device) exchange.getIn().getBody();
NetworkDevice netDevice = new NetworkDevice(device);
exchange.getOut().setBody(netDevice);
```

Now we can create the type converter class and actually use it as a database lookup class:

```
@Converter
public final class DeviceEntityLookup {
    @Converter
    public static Device lookupDevice(NetworkDevice networkDevice ..
```

jeets-camel-netty4-etl

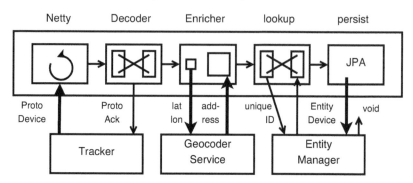

Figure 14.1 At the ETL route start a Netty polling *consumer* is triggered from a tracker client, the enricher *produces* a client request and *consumes* the geocoder response and finally the JPA *produces* the entity to be persisted at the end of the route.

The implementation is 'a walk in the park' and you can step through the class to see how the managed entity is retrieved from the database. There you also have the chance to register unregistered devices or to reject the data.

Note how the device's children position- and event lists retrieved from the database are replaced by the passed in devices children. Another detail you can identify is that `NamedQuerys` are hardcoded in the related entity and not in a separate file, like in Traccar.

We have chosen to lookup the device, i.e. make it managed, and then apply the `.persist` method without returning anything. The reason is that the JPA would drastically slow down a route and we don't want to wait for the managed device returned from the `.merge` method and continue working with it. We'll find a better way to work with incoming device messages without hardware access soon.

14.3   CAMEL GEOCODER COMPONENT

We have created a functional ETL module with a route starting at the Netty component and ending at the JPA component – without really looking into the messages data. We have learned how to create a complete Camel route `from` one endpoint `.to` another – without any data modification. In this chapter we will see how to manipulate data *along* a Camel route and how to direct the routing with Camel's expression language. As the ETL module slows down the route significantly we might as well add another 'slow' component from which to acquire additional data from.

We will look into the messages, extract their geographic coordinates and look up their addresses with the Camel-Google-Geocoder component which is installed by adding the artifact `camel-geocoder` to the ETL's `pom.xml` file. Every Camel component is described on a single html page and at `camel.apache.org/geocoder.html` we can find the simple Geocoder call we are looking for:

```
from("direct:start")
.to("geocoder:latlng:40.714224,-73.961452")
.log("Location ${header.CamelGeocoderAddress}")
```

Without Camel experience this doesn't help very much and we will have to add some more Camel knowledge to apply the geocoder *in* the route. Naturally you should always start by creating a complete route with one `from` and one `.to` endpoint in order to run the module. Then you can modify the route at development time.

Figure 14.1 displays the complete ETL route with access to external resources. Without Camel awareness the route could also be described with client- and server functionalities: First a tracker-client triggers Netty and the Netty consumer, like any Camel consumer, represents a server. Behind the DCS we have added a new icon, the content Enricher EIP, which documents the pattern to obtain data from an external server. The enricher fires a Camel producer, representing a client, and the enricher consumer, i.e. server, receives the enrichment data. While Netty consumes the input and JPA produces the output, the geocoder produces a request and consumes the response inside the route. The geocoder is used as an intermediate component along the route.

14.3.1 Camel's Simple Expression Language

A major problem for Java developers when first confronted with Camel is its explicit terminology for routing with EIPs. We will try to demystify Camel and EIP 'language' by first expressing the solution in Java code and then migrating it to Camel code. Before implementing the above mentioned content enricher we need to learn Camel's expression language called Simple.

So what are we looking for?

We are aware that the DCS component provides a new entity device that can have one or more positions with geographic coordinates. In Java we would simply create a loop to iterate over the positions and look up their address:

```
private void addPositionsAddresses(Device device) {
    for (Position pos : device.getPositions())
        pos.setAddress(
            lookupAddress(pos.getLatitude(), pos.getLongitude()));
}
```

Note that the `void` return does not influence the object references in the Java code. Once you get hold of the device object reference you can lookup each position object referenced on the heap. By setting the position's address the device and its positions have been enriched and don't need to be returned.

For a better representation of the enricher pattern you could also return the device explicitly:

```
private Device addPositionsAddresses(Device device) { .. }
```

Camel should resolve the coupling of the `lookupAddress` method calling the external geocoder service, which can hold up the calling process due to latency and other issues.

The Java method reveals that it is not sufficient to use the device object, which is more than one POJO. It's an ORM with positions (and events) and we need to iterate over the positions, send each coordinate to the geocoder and set the returned address of the position.

Again we'll ask ourselves "What is Camel's core functionality?"

The main actor of the Camel framework is the `Exchange` object, which contains the transfer object traveling along the route. It is created with a consumer request, which is set as the message *body* to be sent along the processing chain. Therefore the Simple language was initially created to look into the `Exchange`. For example you can use Simple expressions to log the current transfer object:

```
from("seda:jeets-dcs")
.log("Received Device Entity '${body.name}' "
    + "with ${body.positions.size} positions.")
```

In this case we know that the `body` from the DCS output is a `Device` object and with `body.name` the Java getter `getName()` is invoked. The second expression goes deeper into the object and references the `Set<Position>` and its `.size()` method.

14.3.2 The Splitter EIP

Now we want to rewrite the loop over the devices positions with Camel Simple. The splitter is the enterprise pattern for iterations – it's a `for` .. `each` loop. With

```
// @formatter:off
.split(simple("${body.positions}"))
  .log("process Position(${body.latitude},${body.longitude})")
  :
.end()
:
```

you create a loop iterating over the devices positions. After the `.end()` of the loop the route continues with the initial device object.

The above route splitter works just like the Java `for..each` loop that we have listed in plain Java. With Camel you can also process all positions in parallel by simply adding a line with `.parallelProcessing()`. Anyhow this is not a good idea for our context as we allow any number of positions in a device and don't want to create the same number of threads internally!

In the listing you will note the `@formatter:off` statement. Its always a good idea, especially in large teams, to work with a code formatter. On the other hand Camel routes are not identified by the formatter and the route gets hard to read. The route is much easier to read when you turn off the formatter to display Camel routes in single lines for each step and add indentations for subroutines.

14.3.3   The Content Enricher EIP

With the splitter we can use the geo coordinates to lookup position addresses from an external geocoder service. And now we can also apply the content enricher to enrich every position with an address:

```
.split(simple("${body.positions}"))
  .enrich("direct:geocode", GeocodeEnricher.setAddress())
  .log("new Address '${body.address}'")
.end()
```

In this listing we are using the `.enrich` DSL method like a submethod to modify the message body. In Camel the submethod relates to a subroute, the first parameter of `.enrich`. Internally the method uses a producer to create positions and their coordinates can be used to invoke the external service for a reply. The second parameter refers to an implementation of an `AggregationStrategy` in which Camel provides the original and the exchange returned from the resource:

```
public Exchange aggregate(Exchange original, Exchange resource) {..}
```

After naming the subroute we need to create it with

```
from("direct:geocode") // body = position
.toD("geocoder:latlng:${body.latitude},${body.longitude}")
.log("Location ${header.CamelGeocoderAddress}");
```

And there you can see how to create a dynamic endpoint with `.toD` and Simple expressions. Now the device's positions have been enriched and the device entity can be sent to the JPA endpoint to be persisted.

14.3.4   Geocoding Strategies

We have created the ETL for use with the Traccar GTS to geocode and persist every incoming GPS message. If you place a debug point inside the `GeocodeEnricher.setAddress()` method and inspect the `GeocodeResponse` you will note that it returns more than 6.000 characters! Why did we go through all the networking with protobuffers to define every single bit and control the overall traffic, if we then stress the network with this massive overhead? For an SDC scenario geocoders should not be used in that way!

Applying a geocoder on every incoming position comes close to spam. Remember that a geocoder returns a human readable address and as long as no one reads it - don't retrieve it. In the Traccar system you could also trigger the geocoder when you gather reports *with individual* positions. Then you could even use the address components to fill in separate columns for the admin hierarchy city, street, country etc. And on a map display you can read the address from the map!

Another interesting aspect of the geocoder is the result type. If you look at the full response you will find

```
GeocoderResult{types=
[bus_station, establishment, point_of_interest, transit_station],
```

which could be used for a dedicated DCS / ETL and port to support transit tracking of a GTFS route to add station names!

14.4 CREATING CAMEL COMPONENTS

In Figure 13.1 on page 142 you can see the complete `jeets-dcs` in EIP notation. Any data is propagated in Camel exchanges up and down the Camel route. Each DCS serves a dedicated protocol, i.e. device protocol to produce a device ORM with system entities. The `jeets-dcs` is ready to be used and developers can add their configuration as needed.

It always makes sense to work with a component for a while to gain experience and add some fine tuning. An important aspect is the use of many `jeets-dcs` for many ports creating different system entities. This will not be covered by the book, but with Traccar you have installed a system with more than a hundred protocol decoders in a single Netty `NioServerSocketChannelFactory`. By looking at the Camel configuration options for Netty you can find options to orchestrate different `jeets-dcss`:

Camel Netty Component [36]

> `bootstrapConfiguration`
> This can be used to reuse the same configuration for multiple consumers, to align their configuration more easily.
> `bossGroup`
> For example to share a thread pool with multiple consumers.
> `workerGroup`
> For example to share a thread pool with multiple consumers or producers.

Of course you are welcome to implement your solution to JEETS or to check the latest implementations after this publication. In any case the component becomes mature by frequent use and different contexts may narrow down its domain use. Then the time has come to model the Camel application to a Camel component. You don't have to wait for a public implementation of a component for your needs. You can create your own!

Wouldn't it be nice to have the DCS as a Camel component where developers and production managers can easily append the configuration something like this:

```
from("jeets-dcs:device://localhost:5200?concurrentConsumers=4")
from("jeets-dcs:device://localhost:5201?concurrentConsumers=2,
                                bootstrap=ServerBootstrap")
from("jeets-dcs:position://localhost:5202")
from("jeets-dcs:geofence://localhost:5203")
from("jeets-etl:device://localhost:5204?geocoder=google,
                                persist=false")
```

Note the component's names `jeets-dcs` and `jeets-etl`! The implementation is hidden to users while the component owner can re/define the configurations. The second parameter after the colon defines any ORM provided in a .`proto` file and refers to a `ServerInitializeFactory` which has to be registered by the protocol name.

14.4.1 Enterprise Development

In the automotive industry IT development is spread over different departments, locations, countries and even subcontractors. Imagine one team responsible for device communication has developed core DCS and ETL with protobuffers, Netty, PU and JPA endpoint just like we have. They can distribute these components as a 'starter kit' that you can use out of the box. Another team is responsible for hardware and provides IP, port and a DBMS to store data. Yet another team has setup the security with authorization and accounting.

Now a new customer, some OEM with a certain car model, wants to develop SDC technologies. A project team is formed and all they have to do is to provide a PU with a matching protocol defined in a `.proto` file. That is our situation right now. We are in the comfortable position to receive device entities and use them for whatever purposes. Naturally these new projects can be started from scratch or, like we did, with copy and paste and then modified for project specifics.

We have now created an ETL module which can be used in conjunction with Traccar to decode, geocode and persist incoming messages. The whole idea was to remodel a tracking system for dedicated purposes by applying JEE components.

V

Middleware

Java Messaging

CONTENTS

15.1 INTRODUCTION

In the last section 'Enterprise Integration' we have learned how to integrate a TCP software bound to the hardware ports with a persistence software to store data in a database. The two core patterns of (or main reasons for) integration are decoupling and threading, which is easy to realize if you were to implement the *complete* ETL route via Request-Reply.

The Ack message would be sent back to the tracker *after* the message has been persisted and the TCP channel would remain open and block resources for a *magnitude* longer! We have decoupled Netty (TCP) from persistence (JPA) to avoid this. Threading was not implemented by us at all – it comes with Netty out of the box. Netty takes care to avoid blocking between different messages arriving from different trackers.

We have achieved the integration with Camel in a single ETL module which is launched via main method from command line. Besides launching the ETL we have implicitly relied on a database management System (DBMS), i.e. PostgeSQL that has to be running and should perform well enough to serve the software load – of many loaders. The administration of these components on one production system is manageable while developers can install and run them on a single PC.

As stated in Section 14.4.1 self driving cars are definitely not developed by a single team at one location. It was also outlined that a single team can be responsible for device communication and provide DCS, ETL and other modules via the enterprise's Maven repository. With this 'starter kit' any team can easily create their own persistence unit and a `proto` file to define the messages for their project context. Enterprise integration patterns become increasingly important for distributed development, since every team always provides some proprietary code and personal style even, if they adhere to the Java enterprise specifications.

JEE specifies the interoperability of applications – beyond Java. For example JPA is the specification to hide the Hibernate implementation and behind it the actual DBMS. By using the `camel-jpa` we can switch the implementation and DBMS according to the enterprise's standards. With the ETL module we have implemented a device communication to collect tracking messages in the database – completely independent of the Traccar system. We can launch Traccar any time later to visualize and analyze the collected data interactively and apply Traccar as a tool.

This separation of executables is especially useful for distributed teams. Next we will explore how we can add middleware to 'buffer' messages according to the consumers load. By setting up a message queue service the DCS module can 'get rid' of the converted messages immediately; even if the next module is too busy to accept messages – or even temporarily offline. We will add a middleware service as a separately installed software component on OS level.

15.2   MESSAGE ORIENTED MIDDLEWARE - MOM

The DCS module to supply system entities is practically useless in a stand alone mode as long as no other module is there to receive, process and propagate this information somewhere, somehow. A device communicator should provide a Camel route to send the bare GPS information to some server without bothering about further processing.

The solution development to this general problem domain started in the 1970s under the term 'Message Oriented Middleware' – software especially for information *exchange*. Distributed applications can communicate via a MOM server in a manner that is asynchronous, loosely-coupled, reliable, scalable, and secure. MOM acts as message mediator between senders and receivers and provides loose coupling among enterprise applications. For a long time MOM solutions were proprietary until the Java Message Service (JMS) was introduced as *the* Standard.

By introducing Camel we have already gotten familiar with the major messaging terms endpoint, consumer, producer etc. We have used the built in SEDA component as DCS endpoint. Problems with SEDA (inside a single Camel context) and the similar VM endpoint (accross different camel contexts) occur, if the JVM terminates during message processing. There is no recovery or persistence mechanism and the messages are lost. By replacing SEDA with an MQ endpoint we gain persistence and reliability of JMS between different Camel applications or more generally between completely different (hardware) systems connected only via an URL implying IP and port. A good start for cloud computing . . .

We will create a DCS module to send the system entities to a message queue where they can reside until one or many applications are ready to consume them.

15.3   JAVA MESSAGING SERVICE - JMS

Middleware software has evolved as a necessity to mediate between data formats, operating systems, protocols and as a welcome side effect between programming languages. The software category of message routing has become a complex discipline of its own. Later the Camel integration framework was invented to simplify routing of messages in a descriptive manner.

JMS is a standardized API to abstract implementations of different MOMs to send and receive information. With the JMS API Java development requires little knowledge of messaging details like transport mechanisms etc. JMS can be viewed as an integration technology itself to allow loosely coupling of applications with reliable message exchange. The messaging is asynchronous by nature as it provides the ability for senders to 'get rid' of their output to a *destination* while the receivers can 'pick up' the messages as appropriate. The middleware does not require sender and receiver to be on line at the same time – they don't even have to know of each other. Beyond notifying applications JMS supports their inter operation.

JMS formalizes the core concepts of messaging and we have already learned some messaging terminology with Camel and it shouldn't come as a big surprise that JMS is based on these terms:

JMS client
An application is written in Java to send (produce) and receive (consume) messages by utilizing the `javax.jms.MessageProducer` and `MessageConsumer` interfaces.

JMS producer
A client application that creates and sends JMS messages.

JMS consumer
A client application that receives and processes JMS messages.

JMS provider
The implementation of the JMS interfaces in Java.
We will use ActiveMQ as the JEETS middleware.

JMS message
The fundamental actor of JMS sent and received by JMS clients.
We have gotten a first impression from Camel's `Exchanges`.

JMS domains
The two styles of messaging – point-to-point and publish/subscribe.

Administered objects
Preconfigured JMS objects that contain provider-specific configuration data for use by clients. These objects are typically accessible by clients via JNDI. JMS spec defines two types of administered objects:

- `ConnectionFactory`
  clients use a connection factory to create connections to the JMS provider.

- `Destination`
  An object to which messages are addressed and sent and from which messages are received.

15.4 ACTIVEMQ

We have chosen the Camel framework to conquer the world of integration and messaging. Camel provides a higher design level to look at and model a system. We have briefly introduced the development from proprietary MOMs to the JMS standard. The terminology seems familiar from Camel and yet *the actual integration technology is messaging* itself!

Historically Camel was created after JMS specified MOMs were on the market. It was initially derived from and optimized for Apache's ActiveMQ implementation [52]. The idea was to create a declarative abstraction level with a simple domain specific language to describe EIPs in a lightweight framework which could be embedded – anywhere.

ActiveMQ is an enterprise message broker, a server always running and available via IP just like a database service. And ActiveMQ is Camel's big brother as you can theoretically do anything (and more) with ActiveMQ that you can do with Camel, only that Camel is descriptive and does not add any technology nor persistence. Camel is not considered to be a middle ware itself as it is mostly concerned with leveraging thirdparty software. We have introduced the world of messaging from the Camel perspective to demonstrate the simplicity of integration with it.

15.4.1 Messaging with ActiveMQ

Apache ActiveMQ is an open source message broker for remote communication between applications and systems via Java message service. ActiveMQ provides a wide range of connectivity with containers, persistence, security and application servers. We will see how to loosely couple different GTS modules with an MQ service and free them from interdependence and timing constraints.

By adding a reliable MQ to our system we can take Camel a step further and make use of (temporary) message persistence *between* different modules. Besides the integration patterns for threading and decoupling the MQ service adds timing independence of different modules which is already very helpful at development time. The MQ becomes the backbone for various interacting JEETS modules and highly improves scalability and configurability of un/required processing.

15.4.2 ActiveMQ Installation and Test

Please read Appendix C.4 and install Active MQ[1]

After installation you can run the example

```
...\apache-activemq-5.15.0\examples\openwire\swissarmy
```

as described in its `readme.md` file:

1. Start ActiveMQ in a console
   Open browser frontend at 127.0.0.1:8161/admin

2. Run ProducerTool
   See console listing for 2000 messages sent

3. Check browser for messages: 127.0.0.1:8161/admin/queues.jsp
   You should see the queue topic `TEST.FOO` with 2000 messages enqueued and pending. Note that the queue is holding the messages although there is no consumer to receive them and the producer can be shut down.

4. Run ConsumerTool
   See console listing for 2000 messages received

5. Check browser for messages dequeued

The examples directory provides a practical start to ActiveMQ and proves that the broker can mediate messages over your network. Of course it makes sense to get familiar with some other examples. You should be aware that this `swissarmy` example is applying plain JMS messaging although producer and consumer sounds familiar from Camel. In the sources of the `Producer-` and `ConsumerTool` you will find the JMS terminology listed earlier on page 165.

Here is a short pseudo listing of the producer:

```
// Create a ConnectionFactory
ActiveMQConnectionFactory connectionFactory = ..

// Create the connection.
```

---

[1]You can install ActiveMQ as a service to boot with the system.

```
connection = connectionFactory.createConnection();
connection.start();

// Create the session
Session session = connection.createSession(...);
if (topic) destination = session.createTopic(subject);
      else destination = session.createQueue(subject);

// Create the producer.
MessageProducer producer = session.createProducer(destination);
if (persistent) producer.setDeliveryMode(PERSISTENT);
         else producer.setDeliveryMode(NON_PERSISTENT);

// send message
producer.send(message);

session.commit();
connection.close();
```

After this jump start you should have a good picture of what's going on and how we can setup an DCS as a message producer. Then the DCS could send its device entities to the MQ without a *consumer* running to receive them! We can create persistent messages for an MQ endpoint and the consumer team/s can pick them up as needed. This functionality exceeds Camel's domain or, on the other hand, ActiveMQ extends Camel with persistent endpoints.

15.5   JeeTS DCS TO ACTIVEMQ

Please stop ActiveMQ, go to the JeeTS repository, open (import) project jeets-dcs-amq, open test class CamelDcsToJmsTest and run testDeviceToJms test case.

This testDeviceToJms test case is written in plain Java to demonstrate sending a device entity to an in queue, and then sending it to another out queue via the simple Camel route

```
from("test-jms:queue:device.in")
// prepare, filter, validate messages
.to("test-jms:queue:device.out");
```

already indicates an overall design model. Any application in the enterprise can pick up the incoming device entities at device.in, process them for a purpose and place the results in any other ActiveMQ endpoint, for example device.out.

For a tracking system you could now add a `Distributer` in the route to filter events and notifications and make them available at the endpoints `device.events` and `device.notifications`. Then, like in Traccar, an `EventManager` and a `NotificationManager` could pick them up and kick off actions (and external resources) configured for dedicated project oriented tracking!

In the `pom.xml` you can find the `activemq-camel` entry which resolves into Camel, Spring and ActiveMQ artifacts and a mixture of `org.apache.camel` and `org.apache.activemq`. The Java code is similar to the the plain JMS code earlier, only simpler as we are working with Camel.

```
// proprietary
ActiveMQConnectionFactory activeMqConnectionFactory
        = new ActiveMQConnectionFactory(...)
// cast to JMS spec
ConnectionFactory connectionFactory = activeMqConnectionFactory;
context.addComponent("test-jms",
    JmsComponent.jmsComponentAutoAcknowledge(connectionFactory));
```

The test can be altered from an external broker to an embedded broker with the `activeMqVmTransport` flag, which makes sense in a Maven environment without external resources. In the end ActiveMQ also is a jar and can be managed via Maven or as in this test be embedded [53] and launched on the fly. By common sense only the official ActiveMQ service is available for all developers and in a production system for all modules.

Then we cast it to a `javax.jms.ConnectionFactory` interface provided by the JMS specification to hide the ActiveMQ implementation. This may be important to meet a company's guideline 'strictly JEE '. Then we pass the factory to Camel's `JmsComponent` and can start using it.

As opposed to the earlier plain JMS example you won't find the creation of a connection, session, topic or queue. It's all in the URI `"test-jms:queue:device.in"`[2]. Even `"test-jms:device.in"` would do the job, since `"queue"` is defaulted to anyway. The rest of the test case applies Camel's convenient `ProducerTemplate` to simulate the DCS.

The test case `testDeviceFromActiveMq` is similar and adds the DCS code created in Chapter 13. Now the slightly modified DCS route `DcsToAmqRoute` looks like this:

```
from("netty4:tcp://localhost:{port}
        ?serverInitializerFactory=#device&sync=true")
.convertBodyTo(Device.class)
.inOnly("activemq:queue:device.in
        ?connectionFactory=#activeMqConnectionFactory")
.process(new AckResponder());
```

---

[2]configuration options at `camel.apache.org/jms.html`

Now we are routing all the way from Camel's `camel-netty4` to the `activemq-camel` component with the URI # registered classes:

```
@Override
protected JndiRegistry createRegistry() throws Exception {
```

and this time we are sending a protobuffer while the the the `"activemq:"` component is receiving a transformed entity!

After you have received a green bar for both tests you can start your local ActiveMQ and manually modify the `activeMqVmTransport` flag to false. The default AMQ credentials are used to connect to the broker:

```
amqFactory = new ActiveMQConnectionFactory(
                ActiveMQConnection.DEFAULT_USER,
                ActiveMQConnection.DEFAULT_PASSWORD,
                ActiveMQConnection.DEFAULT_BROKER_URL);
```

Now you can run the test class and observe the results in the browser `localhost:8161`. Please don't forget to reset the `activeMqVmTransport` flag. The tests are designed to run in the standard build of the JEETS repo which should not require external sources. In a later system design phase the repository will be equipped with Maven profiles to run tests against database, broker etc.

Now you can start ActiveMQ and `jeets-dcs-amq`, open a terminal at `.../jeets-tracker` in the JEETS repository and send the six device samples via command line to ActiveMQ. You should see the six devices in the MQ browser frontend in the queue `device.in`.
Then you can terminate the DCS and ActiveMQ.
After restarting ActiveMQ you should still see the messages in the queue.

Geo Distribution

CONTENTS

16.1   ENTITY ROUTING

By placing the `Device` messages in the queue `device.in` we have reached the end of the *technical* device communication and can begin to *semantically* model our tracking system to process system entities.

Now we can compare Traccar's with the Camel/MQ architecture shown in Figures 12.1 (page 126) and 16.1. Traccar's `Position` entity represents the JEETS `Device` entity and is the result of *all* device communication servers feeding the system. With the new architecture we can scale each module on its own hardware and are not tied to Traccar's monolithic design anymore. Figure 16.1 shows the architecture with independent DCSs that can run on different platforms and send the `Device` message to the ActiveMQ resource (on another host).

Now we can start forgetting the details about protobuffers and look ahead to process entities with any Camel processor required in a project. Different teams can cooperate by providing the JMS endpoint syntax and semantics to each other!

At this point we are leaving the GTS paradigm and face the challenges of connected car technology. The different approach becomes apparent by

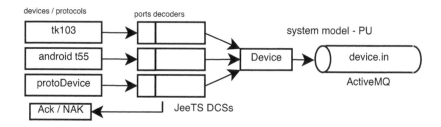

Figure 16.1 Every DCS can run on a different host and send the `Device` message to the ActiveMQ resource to the message queue `device.in`. The DCSs are released and the consumer can pick up the messages as appropriate. Compare to dashed frame in Figure 12.1

comparing a classical GTS geofence manager with the geo router we are about to create:

1. A GPS tracking system is constantly tracking vehicles of a well defined fleet. Geofences can be created for each vehicle and the system can send notifications when this one vehicle enters this exact fence.

2. SDC systems *do not track* single vehicles consequently – actually they never identify vehicles inside the system. The device identity is only used for authentication and authorization on the technical DCS side. An SDC system is receiving 'snippets' of tracks from many different vehicles and can create traffic information with them. Therefore it makes sense to prepare these snippets for different geographical areas, i.e. bound boxes or polygons.

For large projects with cars all over the country the geographical areas could be defined with polygons of the administrative hierarchy retrieved from a digital map with political boundaries[1]. On the other hand the procedure to check a car's location against a complicated admin polygon with many geo coordinates is expensive in terms of machine power.

16.2   THE JeeTS TILE MAPPER

A very effective way to distribute messages is via tiling algorithm. If you work with an online map like Google Maps or OpenStreetMap you will sometimes see the map tiles when you change the zoom level. These squared tiles seamlessly cover a city, country, continent. Zoom levels define precise map scales with an integer number of map tiles covering the world from Greenwich to Greenwich (-180, 0, 180) and from pole to pole (-90, 0, 90) Any position (lat,lon) can be mapped to an exact tile (coordinate x,y) for every zoom level,

---

[1] Have a look at the geocoder response to identify the admin hierarchy.

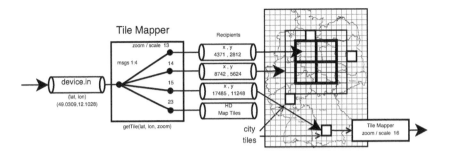

Figure 16.2 The tile mapper can serve as a semantic scaling method to distribute messages to different geographic tiles and different zoom levels, i.e. map scales.

i.e. map scale. By applying this tile algorithm on a car's location the messages can be distributed to different tile servers and, if required, large tile 'containers' can redistribute the cars locations to smaller tiles with larger zoom levels.

The tile algorithm can vary from company to company and should therefore be distributed from a (Maven) repository similar to the the persistence unit for standardization. Algorithms can be provided for different programing languages, as pure math or as a service.

As an example algorithm you can look at the Slippy map tiles [65] function:

```
(lon,lat) to tile numbers
-------------------------
  n = 2 ^ zoom
xtile = n * ((lon_deg + 180) / 360)
ytile = n * (1 - (log(tan(lat_rad) + sec(lat_rad)) / pi)) / 2
```

Although trigonometric calculations can get expensive the formula is convenient for our message distribution. In big data systems it is worthwhile to create tiling algorithms with fixed size integers and without trigonometry to save some more milliseconds.

Let's see how we can distribute messages to tiles with Camel.

You will find a `TileMapper` seed component in the test environment
of the geo router at .../`jeets-server-jse/jeets-geo-router`.
Please run the `AmqTileMapperTest`
before you continue to read about the implementation.

16.2.1   Use Case and Risk

Every position maps to one tile on *every* zoom level – a 1:n relation. Different departments can create completely different applications from the same traffic data flowing in. One department is interested in analyzing traffic flow across the state and request a lower zoom level 5 (0 = world) while the SDC requires accurate traffic information on zoom level 16 (small road).

From this simple use case we can derive that the tile mapper should address different hosts for every tile size, i.e. zoom level, i.e. map scale. And the message will be cloned to every host, i.e. URI.

The following implementation provides a seed component, if you are interested in cloning your messages to different locations. But you should be aware that each host is the start of a completely new and independent system. It would make sense to provide the same message on every map level, if they end up in the same application. Especially at the start of semantic handling, i.e. device.in you should be more concerned with reducing the number of messages and not increasing them. After introducing the tile mapper we will create a geo router to route each message to one destination, i.e. 1:1. The Geo router can be much more effective for fleet management.

16.2.2   The Recipient List EIP

With the Recipient List EIP we can generate a list of recipients, i.e. endpoints based on the message content. The clue is that one or more recipient URIs can be calculated dynamically. Then Camel takes care of forwarding the message to each destination.

A quick start to create a node in a Camel route is to implement a processor:

```
public class TileMapper implements Processor {
  @Override
  public void process(Exchange exchange) throws Exception {
```

With a processor we can receive an exchange, modify it and forward it back to the route. In a next step we copy the tile algorithm from the Openstreetmap Wiki to the method

```
String getTileString(double lat, double lon, int zoom)
```

For our scenario we return a TileString to describe the exact destination endpoint. In the test case we can create destination tiles for different zoom levels

```
(49.03091228,12.10282818) maps to tiles ..
    area      zoom    xTile     yTile     map scale
    village    13      4371      2812      1:70,000
               14      8742      5624      1:35,000
               15     17485     11248      1:15,000
 small road    16     34971     22497      1: 8,000
```

The rest of the implementation is straightforward:

```
// receive Exchange from route
// get LAST Position from Device
// loop over zoom levels (13 to 16)
//   calculate Tile Coordinate for each zoom level
// create String Array as destination list:
//   activemq:z13x4371y2812, activemq:z14x8742y5624
// set destination list in header
// forward Exchange back to route
```

What we learn with the tile mapper is to create or modify exchange headers along the route to set up 'variables' to direct the program flow.

```
from(startEndpointUri)
.process(new TileMapper())
.recipientList(header("tileRecipients"))
.parallelProcessing();
```

The `.recipientList` represents one or more `.to` endpoints, which can be set to `.parallelProcessing()` – with care. When you run the `AmqTileMapperTest` against your ActiveMQ installation you should see these endpoints in the browser frontend:

```
                    Messages
Name               Pending Enqueued Dequeued
tiles.device.in       0       1        1
    z13x4371y2812     1       1        0
    z14x8742y5624     1       1        0
z15x17485y11248       1       1        0
z16x34971y22497       1       1        0
```

We have cloned and distributed one message to four destinations where they can be picked up from 'external' systems. This kind of distribution is useful for isolated 'top secret' projects.

Another application of a tile mapper could be geo scaling. Let's say a small company offers fleet tracking in their area. After a while the growing number of customers is spreading across the country and the existing system is reaching its limits. With a tile mapper you can initially distribute positions to tiles at zoom level 5. Then the backend software could be installed on separate hosts and each host could serve one tile.

On the other hand each Tile URI, i.e. `z13x4371y2812` is constantly collecting messages. Now we can add a monitoring tool and as soon as a single tile reaches a certain threshold another tile mapper is invoked to consume the messages from a level 5 to a number of level 6 tiles. It wouldn't be too complicated to create a frontend with green and red tiles and employees could click on the Red tiles to find more green and red ones.

In the end the tile mapper is a 1:1 function from position to tile for each zoom level and can be applied in many different contexts. In a Camel context we can insert the tile mapper for filtering or distribution as demonstrated.

To keep enterprise software compatible it is more than useful to provide data formats and conversions in tiny jar files from an enterprise Maven repository. The SDC can use many different coordinate systems like GPS, map LinkId with Offset, tile number etc.

16.2.3 Camel POJO Messaging

If you don't like Camel's DSL and prefer to implement a tile mapper with Java only you should look at Camel's POJO messaging example at

> `camel.apache.org/pojo-messaging-example.html`

to find the `DistributeRecordsBean`. The code reveals how to hook up to an endpoint and create the recipient list with annotated methods:

```
public class DistributeRecordsBean {
    @Consume(uri = "activemq:device.in")
    @RecipientList
    public String[] route(String body) {

        // create destination URIs in program logic

        return new String[]
            {"activemq:z13x4371y2812", "activemq:z14x8742y5624"};
    }
}
```

16.2.4 Camel Component String

During the development of some useful JEETS components with the Camel framework we have applied simple strings for components (`"activemq:"`) and endpoints (`"activemq:device.in"`). As a Java programmer it is very simple to create strings quickly. But in a system with a growing number of components you will have to start managing these strings on a system level. Otherwise the messages can appear to travel erratic routes and you might loose control of message cloning, i.e. system load.

Be aware that every component must be well defined and configured in the Camel context. On the other hand this is also a feature. If you think of heavy components like JPA and MQ you will sooner or later add JTA for safe transactions and pooling for better performance. This technical fine tuning can be done *after* modeling the actual routing.

16.3 TRAFFIC MONITOR

The automotive industry consists of OEMs, the actual car vendors, but much more people are working for automotive suppliers, which again rely on additional subcontractors. And with the global target of the self driving car the auto industry has to collaborate with the data hosting industry. Therefore every person in automotive business working on a 'top secret' prototype is also working on the same goal to automate vehicles. For enterprise management this implicit contradiction poses the challenge to separate 'confidential' development from emerging global technologies.

In order to better understand this challenge we can create an application that globally serves a complete enterprise, while hiding the details of a new feature driven by a premium OEM. This customer is spending a lot of money to introduce something to 'his' customers and wouldn't be amused, if another OEM would benefit from his investment. For economic reasons the automotive supplier responsible for the implementation would develop according to company standards and share the know how internally. Synergies are vital for big businesses – just like discretion.

How can we unite and separate software development?

SDC development is a great challenge without a doubt. The car is treated like a child growing up and learning to react to his environment – on his own. In this book we are looking at the connected car, i.e. client server communication, and exploring its limits. The intrinsic limitations are network latency, time-dependent bandwidth, connectivity and positioning precision – besides hardware aspects.

For example GPS tracking *is not* being applied to validate the correct trajectory along the *road width*. In the best case the lane can be *assumed*. A car can be equipped with many onboard sensors and actually 'watch' the road. Client server communication is meant to basically serve the traffic situation around a car. But how can you generate traffic for a certain area at development time?

In Figure 9.2 we have introduced a simple solution by using the `jeets-player` to playback recorded tracks. By combining many GPS players we can generate traffic. But traffic is much too complex to simply play back a few vehicles. Traffic is defined by rush hours, holidays, weekdays and many unforeseeable events like accidents at critical intersections.

From the last section we have outlined that one enterprise is always working on one solution for SDC scenarios – even, if many projects are treated as very confidential relative to external parties. Internally it would be unwise *not* to exchange information between different projects. Therefore it can be helpful to develop a 'neutralized' project to *publish* motion data without any information about the vehicle types.

If you provide GTS services you should never reveal different customers fleets to each other. In the end all customers are ('*horizontal*') competitors in the same market. On the other hand the information streaming in from different fleets can be useful for internal routing and supply added value for everyone. The production team could use the GTS traffic monitor for dynamic load balancing, which is a technical added value.

The most important parameters when setting up a traffic simulation are day and time. A developer can choose a certain area for his simulation and then retrieve the data from the traffic server. This server can offer unix time expressions to define traffic for a certain weekday and aggregate the data for every 'normal' Tuesday around noon. In addition the server could offer some 'special situations' with a previously recorded accident etc. To get a picture you can go to Google maps and turn on the traffic View and choose between live or 'typical' traffic for certain days and times of day.

Usually even large companies use carefully chosen limited locations for elaborate road tests and their developers can simulate the traffic in these locations. In addition to their own test cars the company might acquire additional sources to enrich the (simulated) traffic. For example live data of mobile phone locations can be extremely helpful. For open source development crowd sourcing can form a powerful source and the open data movement is only beginning ... and may become useful in the future to enrich the JEETS .

From the Camel view the idea is to clone the messages from every project, strip them down to the core GPS coordinates lat, lon with a time stamp and send them to some traffic server. This server stores the motion data and every developer can apply the development cycle described in Section 9.3.

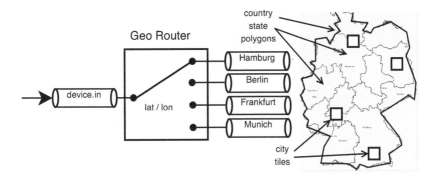

Figure 16.3 With a content (lat,lon) based router the large incoming stream can be reduced, subdivided and directed to dedicated hosts, applications and projects.

16.4 THE JEETS GEO ROUTER

The tile mapper was introduced as a general purpose tool for connected car technologies. We now want to see how we can use a similar distribution strategy for fleet management systems. First of all we don't want to clone messages inside the system at all – too risky. Therefore we should not provide a recipient list implementation to developers to avoid (accidental) cloning.

With the tile mapper implementation we can create 1:n entities. The next implementation of a geo router will provide a 1:1 distribution. To achieve this we will apply the content based router EIP as shown in Figure 16.3. The geo router will check every message and route it to one destination only.

16.4.1 Use Case

For this book we have set up the JEETS repository to develop one system with subsequent testing of all components. Due to the lack of a public 'traffic source' we have created our own public transit (or GTFS) factory in Chapter 11 to generate traffic. We have used real GTFS data from Hamburg's agency HVV (see hvv.de).

By creating a GTFS factory you can generate as much traffic to test your system as needed. You can create a railway vehicle at the exact time it leaves the originating station and simulate its track. For more traffic you could do the same for buses and easily simulate hundreds of vehicles for the city of Hamburg.

Now we will separate this public transit data from the tracking of our production system Traccar. As a long as we are not tracking vehicles in Hamburg this separation of Hamburg from the rest of the world is a feasible development strategy. Business wise the filtering is a bit more complex but has the same effect to direct data streams.

Therefore we will create a geo router to route every device message streaming in from the Hamburg area to the endpoint `"hvv.device.in"`. Any other message will be directed to `"gts.device.in"`. This separation in two streams also separates the course of the book:

1. We will use the `"gts.device.in"` to analyze and plan implementations of classical GTS managers, like in Traccar (see Section 12.3).

2. `"hvv.device.in"` will be picked up in the next part of the book to demonstrate how to send tracking messages to an JEE application server and use them inside an enterprise application (EAR).

16.4.2   The Java Topology Suite - JTS

The core of the tile mapper is a simple formula to map (`lat,lon`) to (`xTile`, `yTile`) which executes fast. For fleet management with no so big data we can use a more expensive formula for more complex mapping with (dynamic) geozones. In order to elegantly work with large geographic polygons we will introduce another small, yet useful and powerful library to execute complex calculations with a simple geometric class model:

The Java Topology Suite - JTS [66]

> "The Java Topology Suite is a Java library for creating and manipulating vector geometry. It also provides a comprehensive set of geometry test cases, and the TestBuilder GUI application for working with and visualizing geometry and JTS functions."

As of version 1.15 JTS is hosted on github and is maintained by the LocationTech [67] working group of the Eclipse Foundation. The packages were renamed from `vividsolutions` (2002-2016) to `org.locationtech.jts..` [68].

Although the topology suite has a rather unspectacular presentation in the Web it is part of many GIS technology stacks. This separation of layers can be described with JTS as pure math, i.e. geometry and trigonometry applied to a GIS with geographical projections.

en.wikipedia.org/wiki/JTS Topology Suite

"Java Topology Suite is an open source Java software library that provides an object model for Euclidean planar linear geometry together with a set of fundamental geometric functions. JTS is primarily intended to be used as a core component of vector-based geomatics software such as geographical information systems.

It can also be used as a general-purpose library providing algorithms in computational geometry."

Beyond plane (and plain) trigonometry JTS implements the geometry model and API defined in the OpenGIS Consortium Simple Features specification for SQL. So we are on a safe path towards real geography. Note that the TileMapper is only pseudo geographic, since it works with pure Java double values for (lat,lon). To get an idea of geographical processing you can have a look at the GeoTools JTS page:

`docs.geotools.org/latest/userguide/library/jts/index.html`

The image displays how the JTA fits into a larger set of tools and the link provides a nice overview and introduction to JTA. We will roll out the `jeets-geo-coder` as a starter to JTS. Once you have added JTS to your projects you will find more and more cases to apply powerful JTS methods. Besides geometric comparisons you can snap any coordinate point to a linestring, calculate areas and distances, determine polygon entry- and exit coordinates and much more...

Please download the Java Topology Suite (v1.15 +)
Start the `..bin/testbuilder` for your OS
Open file `../jeets-geo-router/src/test/resources/U1withGZone.xml`
Open file `Traccar-GBR-U1.png` in the same directory.

The file `U1withGZone.xml` was generated with the HVV GTFS factory for the subway 'U1' with thirteen downtown stations. The image `Traccar-GBR-U1.png` and the geofence were interactively created with the Traccar frontend. You can create your own polygon and pick it up from the `geofences` table. If you compare the two images you can immediately see the two different projections: 2D planar and 3D spherical surface. For JTS this difference does not matter. The two geometries for a subway line and a geozone polygon will be used for testing, pre-development and interactive learning. With the `JtsTest` class you can gain some pure JTS experience.

Note that the `TileMapper` does not have its own project and is implemented

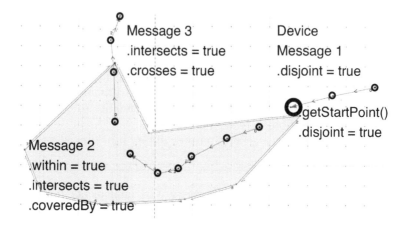

Figure 16.4 The subway U1 is sending three device messages along the trip entering the geozone 'downtown' from east and leaving it towards north. Each message has a different relation to the geozone.

in the geo router's test environment (AmqTileMapperTest). Instead of applying tiling (due to the lack of traffic sources) in the JEETS repository we have chosen to create the jeets-geo-router project which is more useful for classical GPS tracking.

The file U1withGZone.xml represents the complete track of the subway U1 in and through the polygon. In the testbuilder you can open the Predicates screen and hit run to get the Intersection Matrix and Binary Predicates. From a GTS view this track can only be re-constructed from U1 messages along the track. To make the use case more practical we will split the track into three messages as you can see in the file U1splitTrackwithGZone.xml. The latter is only displayed with one track segment in the testbuilder but you can copy and paste the other two into the tool to interactively analyze them while you code. Now you can run the JtsTest to see how to compare the three linestrings with the polygon.

The difference between the two files is also helpful to distinguish GTS and connected car tracking. While the first track can be reassembled and compared to the geozone the latter track consists of three traces along the track which are *not* connected. The three traces were chosen outside, inside and crossing the polygon to provide sufficient data for different cases. Connected car servers are not trying to reconstruct the track in the past and each little trace or mini track is used to extract live traffic data. In an SDC system the three messages are individual 'traffic chunks' and there could be hundreds of other chunks streaming in.

16.4.3   The Content Based Router EIP

The `Device` messages in the queue `device.in` represent entities ready to enter the system. These entities can originate from many different protocols and locations. If you think of a large main frame system for one enterprise you can imagine that this starting point must buffer the incoming device messages for all OEM projects.

In order to deal with the large number of messages in one queue we will create the `jeets-geo-router` project to retrieve these messages and filter, reduce, and distribute them to smaller endpoints for dedicated modules. We will apply the EIP content based router [44] to achieve this *semantic distribution*.

If you compare the Recipient List figure with the (geographic) content based router in Figure 16.3 they look similar with the main difference in 1:n to 1:1 routing of messages. The geo router should separate the Hamburg messages from all others – no additional messages are created.

The `jeets-geo-router` project is easy to build with the Camel experience we have gained so far. First we need to add JTS to the pom file

```
<dependency>
    <groupId>org.locationtech.jts</groupId>
    <artifactId>jts-core</artifactId>
    <version>1.15.0</version>
</dependency>
```

We'll use the same pattern as in the tilemapper by creating a processor:

```
public class GeoBasedRouter implements Processor {

    public GeoBasedRouter(String wktGeometry, String targetTopic) {
        referencePolygon = (Polygon) createJtsGeometry(wktGeometry);
        sendTo = targetTopic;
    }
```

This time we add a constructor to tell the `GeoBasedRouter` to send every message related to the geozone to the `targetTopic`. JTS provides different ways to create JTS geometries and here we are using the Well Known Text format WKT which is an human readable ASCII format – just like the polygons in Traccar!

The second parameter `targetTopic` is something new. So far we have used message queues to send and consume messages to and from the broker. This time we will send messages to Topics. For Camel users this is a tiny change from the URI `"activemq:queue:destination"` to `"activemq:topic:destination"`.

If you have executed the ActiveMQ Swiss Army example in Section 15.4.2 you have already met message Topics. The main idea of Topics is implemented with publish and subscribe semantics. A topic can be subscribed by *all* consumers interested. The consequence for the geo based router is that the route forwards each message to *one* destination, while the consumers represent a 1:n relation of messages. (You could also model n:n.) The advantage is that we don't have to know the number of consumers at construction time. A system architect should be well aware of the difference. Topics can consume many more resources with fewer guarantees while queues are better for load balancing.

The geo router extracts the positions from the device to create a linestring and then 'relates it to the polygon'. As you can see JTS offers many geometric relations and you should carefully choose the one you need. Anyway the predicate matrix represents 'fast' comparisons, but you can find many more ways to compare more details in JTS. For example the linestring also has a direction and we could compare the device track's first point (last point in time) to the polygon with `deviceTrack.getStartPoint()` and decide to only accept messages, if the subway is heading into the zone.

The result of the `GeoBasedRouter` is stored in a header

```
exchange.getIn().setHeader("senddevice", "hvv");
```

The `headerName` "senddevice" is an internal 'variable' for geographic routing. For our simple setting we only distinguish the `headerValue` "senddevice" and "gts".

Now it is easy to create a `GeoBasedRoute` by using the header as a predicate

```
from(startUri)
.process(new GeoBasedRouter(wktPolygon, targetTopic))
.choice()
  .when(header("senddevice").isEqualTo(targetTopic))
    .to("activemq:topic:hvv.device.in")    // JEE
  .otherwise()
    .to("activemq:topic:gts.device.in");   // JSE
```

And that's all there is to it.

JSE Tracking Components

CONTENTS

In this chapter we want to outline the creation of your own dedicated tracking system by combining the technologies we have looked at. After that we will leave Traccar behind and head toward JEE technologies to find out how to inject live data into an application server.

17.1 FROM PIPELINES TO ROUTES

Working against a *production* system is the typical *development* situation. The production system has to be up and running 24/7 to serve customers and generate income and should never be replaced by a new system ad hoc. The new system should rather evolve from the proven production system and replace each module in small steps, continuously. By creating components compatible with the running system new business cases can be addressed while reliable components can remain in place – until resources are available to redesign them for more modern requirements. This approach mitigates the CTOs risk of approving a new (isolated) project with resources (from the production team) and supplies existing interfaces for a new implementation. We have gained an impression by replacing Traccar's DCS with standalone JEETS DCSs.

With respect to the Traccar system there is no need to rewrite the *administration* of devices, people and groups. *Reporting* is applied on the persisted data and the *map frontend* can be used to visualize routes and fleets. JEETS components were rolled out to focus on performance and throughput for a large number of vehicles that exceeds classical fleets. SDC systems are not created to track individual vehicles and can also work with anonymous vehicle data to analyze traffic situations.

The Section 'Enterprise Integration' started with an analysis of Traccar's `Server-`, `DataManager` and `GeocoderHandler` to create an ETL *Camel route* for their consecutive execution. Yet we have not created, nor replaced Traccar's `*Handlers` and `*Managers`. We have only zoomed into the `ServerManager` to look at a single DC server with a single protocol while Traccar serves more than 100 protocols. On the other hand we have created core JEETS components as the building blocks for higher level modules. With many DC servers we can create a `ServerManager` and with various JPA endpoints we can create a `DataManager` and imply a `GeocoderHandler` as needed.

In this chapter we will combine the Spring, Camel and JPA frameworks to demonstrate how you can create your own handlers and managers. As JEE is a very 'tolerant' specification we will have to look at some important details to get the most out of each framework and to avoid 'show stoppers' caused by unspecified coding.

As a guideline we will pick up the brief analysis of the Traccar architecture in Chapter 12.3. Figure 12.1 provides a useful overview for our proceedings. The dashed rectangle represents many DC servers with many different protocols streaming in from many ports of different hardware. The standard entry to a GPS tracking system occurs after all messages are successfully received and transformed into Traccar's system entity `Position`. Then this entity consecutively travels through different *Handlers which are individually communicating with their *Managers.

With Camel we can break out of Traccar's Netty pipeline and create consecutive and/or parallel routes between different (Java-) executables by using Active MQ as an external resource to tie them together. In a distributed system individual components of the production system can be switched off, replaced or cloned while still collecting messages in queues. We can define routes at deployment time by adding or removing components and we can replace standard components with more sophisticated data analysis for premium customers etc.

17.2   TRACCAR HANDLERS AND MANAGERS

Let's have another look at the Traccar managers in the two bottom boxes in Figure 12.1. When Traccar is started via `Main` class the `Context` class creates the managers in its `init` method. The constructors

```
    new GeofenceManager(dataManager);
    new CalendarManager(dataManager);
  new NotificationManager(dataManager);
```

indicate the database connection via `dataManager`. After initialization all managers are accessible via the `Context` class, i.e. `static` access.

The dashed box on the left shows the different DCSs for protocol and port. Each `TrackerServer` is passed to the

```
BasePipelineFactory(TrackerServer, String)
```

where you can find the creation of different *Handlers depending on the configuration properties. Each handler implements

```
protected Map<Event, Position> analyzePosition(Position position)
```

to apply some algorithm on the `Position` to return a newly created `Event`. And finally the events are processed and forwarded by the `NotificationManager`

```
Context.getNotificationManager().updateEvents(events);
```

The `Position` message travels through the handler chain and each handler accesses the required information from the database. Each Traccar *Manager indicates database access.

Now we'll put on our JEE glasses and look at the `EntityManager` specified by JPA. We have already used the entity manager in the `jeets-etl` project in Chapter 14 via Camel's JPA component. For non Camel and experienced JPA developers it was not very satisfactory to retrieve the entity manager from headers inside the transformation class. Camel documentation

[42] states "Its strongly advised to configure the JPA component to use a specific `EntityManagerFactory` instance."

There is much more you can do in terms of performance once you obtain an explicit entity manager. We want to create a tracking component to receive the `Device` ORM (modeled after the system ERM) via a Camel route from some message queue like `"gts.device.in"`.

17.3 HIDING MIDDLEWARE

Camel allows a higher level view on the distributed system for solid component integration. You should now know how to design a Camel route by starting with its `from` and `.to` endpoints. Message transformation can be added to do the data integration between the components' data formats, i.e. from proto to entity. Then you can add message routing and channels according to your requirements. You should have a good picture of what Camel does for you and can gather some ideas by browsing through more than a hundred components already implemented by the Camel riders.

In a large enterprise you should *not* expect every developer to learn Camel just to implement a single processing step. You should always distinguish between a Camel developer or architect and a Java developer who is implementing a processing step of his expertise inside a Camel route.

The 'integration technique' to achieve this is described on the Camel website as 'Hiding Middleware'. The idea is to provide a Java skeleton class that can be wired to a Camel route, i.e. has `from` and `.to` endpoints. The developer responsible for the Camel route can then define the two endpoints and hide the middleware connectors to a JMS, JPA and many other components and resources at development time. With this role separation business logic, performance tuning and middleware code can be decoupled to a high degree.

How to decouple from middleware APIs [45]

"...you can choose the right middleware solution for your deployment and switch at any time you don't have to spend a large amount of time learning the specifics of any particular technology...

...we recommend you use Camel annotations to bind your services and business logic to Camel components which means you can then easily switch [the URIs and resources]...

Another approach is to bind Java beans to Camel endpoints via the Bean integration. For example using POJO Consuming and POJO Producing you can avoid using any Camel APIs to decouple your code both from middleware APIs and Camel APIs!"

Please mvn install the jeets-db-managers project
and open it in your IDE to follow the proceedings.

17.3.1 Camel POJO Consumer

The HandlerRoute class should represent Traccar's *complete* handler pipeline,
i.e. BasePipeline. The following step by step instructions demonstrate how to
create a Java skeleton class. Imagine one developer is responsible for Traccar's
GeofenceManager with domain knowledge of map primitives (points, Lines,
polygons, etc.) and geometric algorithms. This developer is also in charge of
adding new geofencing features for an impatient customer. The business is in
the situation to create a new system over a longer term, while the customer
is too valuable to have him wait for the new system. The geo developer can
then implement the geofencing feature for the production system. Once the
feature is successfully released the architect of the new system provides a Java
skeleton and the geo developer can transfer the code from the production- to
the new system.

With Camel POJO Consuming [46] you can setup a plain Java class and
simply annotate any method with

```
@Consume(uri = "gts.device.in")
public String[] consume(Exchange exchange) { ..
```

and as we are using Spring the class is registered as a bean

```
<bean id="HandlerRoute" class="org.jeets.managers.HandlerRoute" />
```

With this bean binding [47] the inbound message is then converted to the
parameter list used to invoke the method. Internally Camel creates the route

```
from("gts.device.in").bean(HandlerRoute, "consume");
```

Now we have defined the input from the Camel endpoint. In the method
parameter you can define what exactly you want to receive and we'll choose
the Camel Exchange where we can pick up the incoming device message with

```
Device inDevice = (Device) exchange.getIn().getBody();
```

can set the outgoing message (return value) with

```
exchange.getIn().setBody(gtsDevice);
```

and optionally pass some headers for consecutive consumers (of other departments). This is a very general approach for a seed component that can be customized to your needs. To make it even more general we'll annotate the method with `@RecipientList` to define zero, one or more destinations after going through different handling of the message. Practically we can now modify the message (and headers), set it on the exchange and add a code line

```
return new String[] {"activemq:notify", "activemq:alarm"};
```

anywhere in the program flow to continue the Camel route to *wherever* – hidden from and irrelevant for the component developer. We want to code against the database to retrieve and persist data. As (implicit) result we could also end the route by returning an empty `new String[]{}` array. With input and output we're all set to create a handler route inside the `HandlerRoute` class.

17.4   ENTITY MANAGEMENT

In Chapter 6 we defined the `jeets-pu-traccar-jpa` in a very early design phase as *the* persistence unit of the target system Traccar. The term 'persistence' relates to DBMSs and is specified by JPA. As stated earlier this terminology is a little misleading and narrows it down to database operations. The Java POJOs of the PU represent entities related in an ORM. A more pragmatic name than PU would be 'system entities' to exclude the term persistence. Remember that many connected car technologies can work without persisting entities and recall that the `jeets-dcs` creates device ORMs without any database access and this device can travel down a pipeline / route for evaluation / modification of its content.

The Camel integration framework was created to connect different technologies inside one or more routes, while the JPA framework was created to synchronize `@Entity`-s inside the Java software with their persisted counter part in the database. Therefore the JPA component is simply a node in the route and we had to apply a transformer to obtain the `EntityManager` from the exchange headers. The challenge is that the JPA does not know of the external route, vice versa by injecting a `@PersistenceContext` into a (Spring) bean Camel does not know of it.

If you want to use a single entity manager along a Camel route you will have to create your own mechanism to achieve this. One example is 'The Camel entity manager Post processor' by Flemming Harms at

```
fharms.github.io/apachecamel/2016/11/26
    /Take-JPA-to-another-level-in-Apache-Camel
```

Anyway we will stick to the clear separation of Camel routes and JPA nodes. Our JPA processing logic takes place inside the `HandlerRoute` and we will create a supporting data access object `ManagersDao` to trigger the actual database interaction. With the `HandlerRoute` class we want to model a chain of handlers which can enhance incoming messages with database information. Along this chain we want to retrieve persisted data inside a single @PersistenceContext and let the EntityManager decide when to actually access the hardware resources and when to buffer messages instead of persisting them immediately. This is not only a technical, but also a semantic challenge to retrieve data in the correct order or by explicit queries.

17.4.1 Persistence Context

While the HandlerRoute class represents the Traccar handlers box in Figure 12.1, the ManagersDao class should represent the Traccar managers box. The data Access object -Dao is responsible for all database interactions and the code should work with the same EntityManager instance. Now IoC comes into play and with the injection

```
@PersistenceContext
EntityManager em;
```

we will a get our entity manager automatically. And even better the em can be provided with jeets-pu-traccar-jpa which is configured for the Traccar ERM database on Postgres. The persistence unit is added to the POM and the configuration for Spring, Camel and JPA resides in the file spring-camel-jpa.xml with the major components:

```
<bean id="ManagersDao" class="org.jeets.managers.ManagersDao" />

<!-- access to EntityManagerFactory/EntityManager -->
<bean class="org.springframework.orm.jpa.support.
              PersistenceAnnotationBeanPostProcessor" />

<bean id="entityManagerFactory" class="org.springframework.
                orm.jpa.LocalEntityManagerFactoryBean">
  <property name="persistenceUnitName" value="jeets-pu-traccar-jpa" />
</bean>
```

Figure 17.1 provides an overview for the handler component. The plain Java consumer can be configured for any Camel producer and can optionally send messages to other consumers, i.e. recipients. Internally the handler route is supported by the manager DAO which is coded with a single entity manager.

Developers used to programming against JDBC with SQL statements should spend some time looking at vital design patterns in order to profit from JPA. When designing one DAO you should always keep its semantics in mind. In our case the incoming message is synchronized with the database to check whether the device ID is registered. If the device is found the DAO returns the device message from the database. Then the programmer has two device messages to work with: one incoming device message and one retrieved from the database and attached to the entity manager.

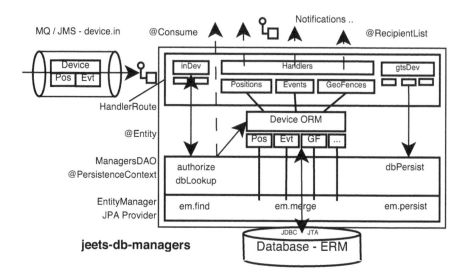

Figure 17.1 The HandlerRoute class @Consumes exchanges from a Camel route and @Produces optional recipient URIs. Internally the each handler accesses the DAO class to synchronize messages with the database. Once the message is attached to the EntityManager each handler can access device information directly from the @Entity.

To get the most out of JPA entity management every manager along the route should respect the operations of previous managers and perceive the handler- and manager chain as one process on one message. This message is validated, maybe modified and finally synchronized with the database. If another route with a different persistence context is processing the same message then the synchronized data *can* become available along each route.

Figure 17.1 demonstrates that the database ORM is retrieved *once* at the beginning of the handler route. Then the code forming the route is programmed against the 'managed' device ORM from the database and is 'attached' to the entity manager. Only at the end of the route there is a second and *final* entity manager call to persist the programmatically modified ORM. By using `em.merge()` you can achieve the same effect and receive a return ORM which can be sent on another route.

Let's look at the important difference of JPA to the SQL programming applied in different Traccar managers. First we have to recall that we are working with three system models! We have designed the relational proto message (1) and the persistence unit (2) after the Traccar ERM (3). As described in Chapter 5 each model uses *almost* the same entities. The protocol messages add some network attributes and the JEETS PU only carries a fraction of the ERM.

17.4.2   Extending PU and ORM

Now we will extend the PU as described in Section 5.6. In order to keep up with Traccar changes to the ERM the PU was initially designed with the three entities `Device`, `Position/s` and `Event/s` to be able to persist messages. To apply an entity manager we will add geofencing to the PU model.

Note that these geofences *can not* be transferred over the network without extending the protocol model. The project lead should pay careful attention not to distribute the protocol package to backend developers dealing with the PU. Both packages have most entities and fields in common and can easily be confused when the developer uses code completion with the wrong project. The PU can be extended to serve every manager analog to Traccar managers.

First we add the entities, i.e. tables to the `persistence.xml` file:

```
<class>org.jeets.model.traccar.jpa.DeviceGeofence</class>
<class>org.jeets.model.traccar.jpa.DeviceGeofenceId</class>
<class>org.jeets.model.traccar.jpa.Geofence</class>
```

then we need to extend the device entity class with a Set of geofences:

```
private Set<DeviceGeofence> deviceGeofences
              = new HashSet<DeviceGeofence>(0);
```

17.4.3   Persistence Tuning

By creating database tables and Java entities the architect can implement JPA component skeletons with the standard JPA API. During production time the database will increase while performance decreases. Production operators should always keep an eye on database performance and archive log files for benchmarking 'standard behavior'.

For JEE component prototyping you can start out with little code according to specification. Initially we created a simple PU from three entities, then we added some annotations for cascaded persists etc. Development and testing works fine with the generated entities and we can send messages with positions to verify the functionality.

One of the greatest efforts of JPA was to unite database modeling and object development. You can traverse an entities Collection of Children objects in a simple loop to retrieve them from the database. But when the first components are released to the production system you can run into performance issues and need to look under the hood.

The main idea of the JEE specification is *not* to depend on implementations of vendors and as a company guideline JPA modules should run in any JEE specified system. This is great for development, but when it comes to deployment any change of the JPA implementation or database vendor can lead to different runtimes, raise timing issues of different routes etc.

In order to improve database Access you can analyze and fine tune each of the following software layers:

1. Java entity code with JPA annotations
   designed after RDBMS modeling practice

2. DAO code with `@PersistenceContext` and `EntityManager`

3. Different JPA implementations
   with different internals to access the database.

4. JPA generation of different SQL dialects of the target RDBMS.

The first two are under <u>Deve</u>loper control, while the latter belong to release management, configuration... <u>Operations</u>. In modern development jargon all layers are under *DevOps* control.

After introducing the `Set<DeviceGeofence>` member to the `Device` entity we can add the Relation

```
@OneToMany(fetch = FetchType.LAZY, mappedBy = "device",
           cascade = CascadeType.PERSIST)
public Set<DeviceGeofence> getDeviceGeofences() {
    return this.deviceGeofences;
}
```

Relations are fundamental to object oriented design and define collaborations to solve realistic problems. They take place in data structures and in software processes – modeled with JPA.

Whenever you extend the persistence unit with new relations you should carefully evaluate every combination of entities and entity collections. You should be aware that most tracking components are actually operating on a single message. When the system receives one device entity the complete software is operating on this very device. If you open paths through the complete ERM you might provide navigation to a different device via `@OneToMany` relation to the devices table. This is potentially harmful und mostly unwanted for a handler route designed for a single device. You should also carefully consider modeling object relations for only one (uni) or two (bi) directions.

Why is the Hibernate entity generation modeling child collections with a Java `Set<>`? Because a `Set` is a very general `Collection` with the same API and should be perceived as a *SQL result set*. The relation can also be modeled with the `List` or `Map` interfaces of the `java.util` package. With a `Set` the JPA implementation is most generally abstracted and can be replaced by a JPA provider specific type. If the internal implementation transforms the `Set` to a `List` then the returned collection will be transformed back and the `Set` is actually holding a well defined `List`.

Note that the entities children are fetched LAZY. In the case of geofences we don't expect too many for a single device, but in the case of positions we must expect any number. Therefore we instruct the entity manager to allocate children entities *without* directly retrieving them. The entity manager is then aware of the Set of children and can decide when to retrieve them and how many at a time.

17.4.4 Ordered Relations

The Java developer can comfortably traverse the device's positions in a loop

```
for (Position position : dbDevice.getPositions()) {
```

but does *not* control how and when the Set<Position> is loaded; how many Positions, in what order? Coders should always be aware of the loading mechanism and can take purely logical aspects into account: Imagine you receive a message from a device that has been registered for a long period. If you invoke the size method PersistedSet.size() it is only logical that the entity manager has to check the database for *all* positions somehow.

To get some more control over the position retrieval we have tuned the Device entities positions from a Set to a List and added *reverse chronological* ordering statements as database hints:

```
@OneToMany(...)
@OrderBy ("fixtime DESC, devicetime DESC, servertime DESC")
public List<Position> getPositions() {
```

This sorting *does* cost performance in the RDBMS, but as all messages should arrive (more or less) in chronological order it is reasonable to set a *standard* order directly in the parent entity. This does not only apply to the handler route and persistence context. It applies to the entities children and reflects the reverse order of the data collection and persisting.

Every time the list of positions is accessed from the database, the database engine does the ordering before it returns the first entries. The order of chronological tracking is practically reversed. The first position in the list is the last position received in the database! The most important part of any PU modification is to spread release notes among JPA developers!

Therefore every Manager implementation should traverse the list *into the past* while the incoming messages list their positions in correct chronological order. If every manager would revert the lists for better code readability then this would consume quite some overall performance in software *and* database! As long as the system is running stable you may not even notice it, but as the database grows it can become a bottle neck. On the other hand, sorting a sorted list doesn't cost much additional processing and can stay in the code as an assertion of the correct order. Anyway the operator should keep an eye on the entity handling route.

Please open the `SpringJpaTrxTest.testSpringJpaTrx` test. This test is using the original PU specified in the `pom.xml` and in the `spring-camel-jpa.xml` files and should connect to your Traccar database. This is useful as you can work with devices having *many* positions to experience loading delays. You can override the PU property `showSql` in the Spring file to analyze JPAs internal SQL statements.

For a practical approach you should look up a device's `uniqueId` with many positions in your database and modify the code accordingly. Open the `SpringJpaTrxTest.testSpringJpaTrx` in your IDE and place a break point at

```
public Device authorizeDevice(Device inDevice) {
    Device dbDevice = null;
    try {
        dbDevice = dBlookup(inDevice.getUniqueid());  // breakpoint
```

step over this line in debug mode and inspect the `dbDevice` to find

```
dbDevice = Device (id=78)
    :
deviceGeofences = PersistedSet (id=87)
         events = PersistedSet (id=92)
      positions = PersistedBag (id=108)
    :
```

that the children (modeled in the PU's entities) have been allocated in a `PersistedSet` or `-Bag`. These collections have been loaded LAZY and the JPA does not define their status at this point. Each JPA provider has its own approach to load collections when they are requested. If you arrange the debugger in a way that you can also watch the console output you can actually see the SQL invocation when you click on one of the `Persisted-` collections.

17.4.5 ORM Navigation

At the start of the `HandlerRoute` the device message is cast from the exchanges body. This way we can still access the exchange and retrieve headers from the exchange to direct the code in more detail. Then the message is used to look up the device entity from the database via DAO

```
Device gtsDevice = dao.authorizeDevice(inDevice);
```

The authorization method, a vital piece of *business code*, creates a named query to lookup and return the managed entity. If it doesn't exist you could create your own business code to send all unregistered devices to an endpoint

```
return new String[] { "activemq:unregisteredDevices" };
```

and / or persist the new device. At this point the database synchronization has started and the programmer can use the incoming message and the database entity for his business logic. Using the JPA loop

```
for (Position position : dbDevice.getPositions()) {
    //  access position entity <- load from database ?
```

is convenient and the code is very readable.

Internally the retrieval of related objects can sum up to many SQL statements on different connection threads to the database. LAZY loading can also lead to performance problems, if the loop invokes a single SQL statement in each cycle and for every position and raise the 'n+1 select' problem that you can find in many books. Since LAZY loading is only specified as a hint to the JPA implementation you should always check the generated SQL statements.

As an example we'll elaborate a theoretical `TripManager`. Trip management should not be under estimated in its performance requirements. Especially in live vehicle tracking we don't want the system to track the complete history of a device from the day it was registered. On the other hand we do need at least one position from the database in order to analyze the route from the last known position to the first position of a new message.

In Traccar the database is accessed from the `DataManager` by the method `getLatestPositions()`. With the `ManagersDao` we can simply retrieve the first position, being the last persisted position, with

```
Position latestPosition = gtsDevice.getPositions().get(0)
```

– without creating another explicit query! The entity manager will take care of it implicitly.

How far back do we have to go in the history of a device? If you have business customers creating official trip reports you are also aware of the complexity of a trip. The simplest trip is the route from home to customer and back home. But what if you stop for a business lunch along the trip? What if the trip lasts for two days? All of these rules are coded from the customer requirements and generating trip reports can significantly slow down the system. After all each trip generation *has to* go through all positions in the database.

Camel can help you send trip processing to a parallel route and continue track processing in the faster handler route. Or if your system is scaled well to track every vehicle and you detect runtime resources you may want to move the determination of a trip into the handler route. Once you acquire the first position of a trip the advantage is that you can load all positions of this trip to the RAM. Then the process only has to determine the end of a trip along the route and will consume fewer resources.

17.4.6   Query Languages SQL and JPQL

Let's see how we can load exactly one trip and leave it to the entity manager to keep it in its cache. In the best case the entity manager has the trip available when the next message arrives from the same vehicle. This assumption is reasonable for a manageable number of vehicles in a fleet and works well – if you set up an GTS instance for each customer, the Docker approach.

For a theoretical discussion we will assume that the device entity is holding the first `Position` of a trip or `null` for a new registered device, just as it holds the last submitted position already. But we can't modify the production system. . .

Note that this 'tripStart' field can only be set when a new message arrives that does not meet the criteria to belong to the trip. It could be added to the database by some background process (i.e. `tripTimeout` etc.) – on a separate Camel route. With an additional trip attribute you could even display the duration in a dashboard.

Experienced database programmers tend to apply native SQL to get what they want. But then they (implicitly) take over responsibility for the generation of managed entities and their attributes for an application. As a general hint you should try to use the JPQL standard as much as possible when you work with the JPA specification. This way you leave it to the JPA provider to generate the SQL code for different database platforms.

With three methods you can create a `Query` instance from the `EntityManger`

1. `createNamedQuery()`

2. `createNativeQuery()`

3. `createQuery()`

Later the interface type `TypedQuery` was added for type safety and should be preferred to explicit casting, especially for entity collections:

```
public <T> TypedQuery<T>       // T = Entity Type from PU
          createQuery ( String name, Class <T> resultClass );
public <T> TypedQuery<T>
      createNamedQuery ( String name, Class <T> resultClass );
```

For trip calculations we don't want to access the device's positions and implicitly trigger some undefined LAZY loading mechanism. To avoid this we can unset the positions' persisted bag of the retrieved device with:

```
gtsDevice.setPositions(null)
```

The point here is that the Collection itself is not attached to the entity manager and even if some positions have been loaded with IDs already they are accessible in the persistence context – even when accessing them from a different method in a different context. Now we will create a DAO method to completely load a well defined sequence of positions of a trip. The method requires two parameters from the device

```
List<Position> tripPositions = dao.loadPositions
    (gtsDevice.getUniqueid(), gtsDevice.getFirstTripPosition())
```

to form some SQL statement. The `firstTripPosition` should be the first position defined by a Trip and we want to retrieve the remaining positions from the database. We have already used simple JPQL to retrieve a device

```
Device device = em
    .createNamedQuery("findDeviceByUniqueId", Device.class)
    .setParameter("uniqueid", uniqueId)
    .getSingleResult();
```

with the `NamedQuery`

```
"SELECT d FROM Device d WHERE d.uniqueid = :uniqueid"
```

that looks like SQL. The difference is that the selection actually defines an entity that must be familiar to the entity manager. The following method applies JPQL with type Safety to create and load the list

```
public List<Position> loadPositions
            (final String uniqueId, Position fromPosition) {

    TypedQuery<Position> query = em
        .createQuery("select distinct p from Position p, Device d "
                + "where p.device.uniqueid=:uniqueid "
                + " and p.fixtime>=:fromdate "
                + "order by p.fixtime ", Position.class)
//    .setFirstResult(100)      // offset
//    .setMaxResults(10)        // pagination
        .setParameter("uniqueid", uniqueId)
        .setParameter("fromdate", fromPosition.getFixtime());

    return query.getResultList();
}
```

First we define the return type for single results or a result list. Then we `select` the entities we are interested in. Note that the selected entities are not explicitly joined! The relation is already defined in the entities Java code as implicit join. Therefore we can use the relation `p.device` in the where clause and we can define a correct chronological order explicitly, since we are not loading the devices children via the encoded reverse ordered relation. The method `getResultList()` of the `TypedQuery` interface returns the complete list of entities with built in type casting. The select statement to load positions includes two implicitly related entities and two where clauses for each one.

JDBC programmers are used to managing the number of loaded records in the code. For example some 'paging' mechanism is usually applied, if a user wants to display large lists in the browser. In the above listing you can find the two parameters to apply in your code

```
.setFirstResult(100)    // offset
.setMaxResults(10)      // pagination
```

For very large trips you can adjust your code to load portions of the trip by using offset and pagination values in the process to avoid the instantiation of too many objects in the RAM. The advantage of using JPQL is that the JPA implementation translates these methods into the specific database dialects:

H2 and Postgres: `limit 10 offset 100`

MySQL: `limit 100, 10`

Oracle creates SQL statements
with subselects and its proprietary `rownum`

This is only a simple example of how one JPA implementation creates individual SQL statements for different platforms. If you would code `limit` and `offset` in native SQL you'd run into syntactic problems when you switch from one database platform to another. The switch from H2 (testing) to Postgres (developer data) would work as they use the same syntax but then the production environment uses an Oracle database and your are in trouble.

Lesson to learn: Use as many JPQL queries as possible and leave the generation of complex SQL statements to the JPA implementation since the syntax can highly differ for different targets. Keep an eye on the runtime behavior when switching platforms and analyze the generated SQL statements at development time.

This book is not about the specifics of JPA and you should spend some time to get used to JPQL. With JPQL you can create left joins, fetch joins to deactivate LAZY loading, subselects with functions and aggregations etc. You should also have a look at an entity relation modeled with a Java `Map<>` and consider it for dedicated use cases...

17.4.7   End of Persistence Context

Figure 17.1 is only indicating a handler route and the reason to place it in one project with the DAO is to keep control of the database synchronization. On the other hand you can create `Managers` as pure Java modules standardized with system entities as arguments. As a manager's complexity grows you can shift it to its own Maven project as a reusable module for another context, even without Camel.

Along the route the entity *can* be used to retrieve related entities. The processing steps should follow the logical order to retrieve entities and analyze positions, then events, followed by geofencing parallel to trip analysis etc.

At the end of the route the entity manager can be filled with positions from LAZY loading and explicit loading. Every modified entity is marked 'dirty' and is triggered by each persist. The order of positions doesn't matter (only the internal IDs) and you can implicitly add (or unite with UNION) every position to the device to trigger cascaded persists. With

```
gtsDevice.getPositions().addAll(inDevice.getPositions());
```

you can combine database positions with message positions. Just before persisting the position order is irrelevant. Later you might add another list from trip management with Java Collection 'algebra'. Then you end the persistence context with the method

```
@Transactional
public void dBpersist(Device gtsDevice) {
    em.persist(gtsDevice);
}
```

You should also get familiar with `em.merge` and `em.flush` to get the best performance with JPA and find out *when* to clean the caches and when *not* to. The first level cache is the (invisible) layer between database engine and software. For tracking a reasonable number of vehicles it *can* be a mistake to flush managed entities. Then they will have to be retrieved again with every new message!

17.4.8 Persistence and Transactions

After having a closer look at JPA and persistence you might wonder why we haven't mentioned transactions at all. For one thing the methods `getResultList()` and `getSingleResult()` do not require transaction handling. If they are invoked inside another transaction, the entities are managed as long as the transaction lasts. Only after commit or rollback the entities are detached.

The reason not to go into the details of transaction management is that we are also using Spring for our database module. You may have noticed the `@Transactional` annotation on the `dbPersist` method. That is all you have to do to wrap the method in a transaction. No more `joinTransaction()` programming etc.

The managers DAO was created for interaction with the database and is provided with a `@PersistenceContext` by IoC. Different JPA implementations can apply additional rules or require special handling to explicitly specify the behavior you expect. In the case of Hibernate every public method is handled inside an explicit persistence context.

The method

```
public Device dbLookup(final String uniqueId) throws Exception {
```

looks up a single device in the database. But when the method returns the device it loses the connection to the entity manager. This strict behavior can be required by company guidelines (atomic units) but as we are coding our private module we can make the persistence context less strict by extending it with

```
@PersistenceContext(type = PersistenceContextType.EXTENDED)
EntityManager em;
```

This way we can handle the device entity inside the handler route and outside the DAO.

Besides Camel and JPA we can kick in the Spring framework easily to take care of transactions. To configure Spring transactions we wrap the entity manager factory in an JPA transaction manager and add annotation scanning in spring-camel-jpa.xml:

```
<bean id="transactionManager" class="org.springframework.
                          orm.jpa.JpaTransactionManager">
  <property name="entityManagerFactory" ref="entityManagerFactory" />
</bean>
<tx:annotation-driven transaction-manager="transactionManager"/>

<bean id="transactionTemplate" class="org.springframework.
                          transaction.support.TransactionTemplate">
  <property name="transactionManager" ref="transactionManager" />
</bean>
```

We have now designed a handler route to retrieve data at the beginning and then work with the returned entity. The 'entry point' to managed entities must be modeled with SQL and JPQL statements. With the message- and database information for one device we have access to all of its positions, events and geofences! Once the information is available many modules can operate on the devices tracks, filter and modify them and reattach them to the ORM.

Children sets can be added to the PU for every new module and the responsible PU coder has to create additional tests that can serve as a code template. Since the PU is the systems fundament every change *must* be tested for the complete repository before it is released.

17.5   GEOFENCE MANAGER

After modeling the route for database management we'll have another look at the JEETS architecture. Compared to our 'production system Traccar' we have seen how to isolate the complete device communication and how we can pick up the entities from a message queue. The messages from the MQ are directed to the handler route to manage all database accesses in a logical sequence. With this route we have created the backbone for a complete GTS and we only have to add event handling and a notification engine.

As a start we now will add the geofence manager as an implementation example and we will look at the logical processing from one manager to the next. Generally speaking the handler route provides the two entities inDevice with chronological positions and dbDevice with reversed chronological positions. The latter is the target entity to be persisted at the end of the route.

```
dbDevice (persisted data)
+--<  position    0          1          2     <--- direction of travel
|     fixtimes 00:41:29  00:39:29  00:37:29     reverse chrono order
|     inDevice (incoming message)
+-->  position    0          1          2     ---> direction of travel
      fixtimes 00:42:29  00:43:29  00:45:29     chronological order
```

We could merge the two entities into the target entity from the start. But to keep managers as independent from the sequence as possible we'll make a contract to keep both entities for all managers. This way every manager can distinguish existing from incoming data. Since the device is part of the system model every manager is implicitly standardized to the system and can be applied in various plain Java contexts.

The device entity is ready to traverse the handler route and every programmer can add children entities from his analysis. We have actually created a new structure to implement Traccar's abstract method:

```
Map<Event, Position> analyzePosition(Position position)
```

If you decide to code modules to retrieve *all* children positions you can maintain performance with automated database archiving. A daemon can pick up all positions after a week or month and aggregate them into a report structure. This way the customer still has access to reports, but can not view the detailed route anymore. These overall aspects of a system are typical for SDC scenarios and even in classical GTS tracking you should define and communicate for how long single positions are persisted. Vice versa the customer might ask you to delete the data due to legal issues. Legal and business aspects are always part of the system architecture.

Please open and run the `GeofenceManagerTest`.
Note that this test is a plain Java and JUnit test
and does not apply any Spring nor Camel!
The test case uses the track and geofence data
depicted on Figure 16.4, page 182.

17.5.1   Implementation

Now we are well prepared to add a pure Java `GeofenceManager` to the handler
route via IoC

```
@BeanInject("GeofenceManager")
private GeofenceManager geofenceManager;
```

and we'll keep an eye on the `org.traccar.database.GeofenceManager` implementation as our guideline. For a start we'll pass the incoming device message
and the managed device entity as method parameters:

```
public void analyzeGeofences(Device inDevice, Device gtsDevice) {
```

The first thing to note is the missing return value. The advantage of working
with managed entities (or generally object references) is that you can modify
the `gtsDevice` without entity manager and the next manager can continue
to analyze and modify it along the route. For example you could analyze
incoming events with a dedicated `EventManager`

```
public void analyzeEvents(Device inDevice, Device gtsDevice) {
```

The demonstration of extending the PU to include geofences can be applied for any manager specific information and the system entities provide the
system dataformats.

Each Traccar manager is implicitly taking permission management into
account:

```
public class GeofenceManager
        extends ExtendedObjectManager<Geofence> {
    :
    @Override
    public final void refreshExtendedPermissions() {
        super.refreshExtendedPermissions();
        recalculateDevicesGeofences();
    }
```

Traccar internally accesses the `ExtendedObjectManager` to query `Group`- and `Device` items and check their permissions in the database. The useful aspect is that all device, user and group permissions are modeled in tables of the ruling ERM. Therefore we could create more complex SQL statements at the start of a route to respect permissions – without modifying the Java code. If the device or user is *not permitted* to access the system the SQL statements should return an *empty set* and the route can terminate this process as a fail fast mechanism.

The test case is based on the data shown in Figure 16.4 and submits the same three portions of the track. These parts have to be combined in the manager logic. As we want to test the system entities without database access we will provide a method to simulate the internals of a JPA implementation:

```
Device dbDevice = simulateDatabaseLookup(devices.get(0));
```

If you look at the method you will see how a geofence is created and related to the device with the `m:n` `DeviceGeofence` entity. To achieve the same effect with a database lookup you should look at your device, pick a part of its trace and create a geofence with the Traccar frontend. The test case also ensures the reverse chronological order of positions to simulate the ordered lookup of the device's position children.

The `dbDevice` is prepared to represent the first part of the track as preexisting in the database. The second part is loaded to the `inDevice` to represent the arriving message. This time the positions are in the correct chronological order. And there we go into the manager implementation[1]

```
GeofenceManager gfManager = new GeofenceManager();
gfManager.analyzeGeofences(inDevice, dbDevice);
```

The geofence manager loops through all geofences attached to the device and passes them to the submethod with all parameters:

```
analyzeGeofence
    (Device dbDevice, Device msgDevice, Geofence geofence) {
```

In this method we can create the `wkt` geometries for JTS analysis.
We can traverse the route from the last persisted position

```
Position lastPosition = dbDevice.getPositions().get(0);
```

through all new positions of the message and check if the edge[2] crosses the geofence

```
if ( jtsLine.crosses(jtsPolygon) )
    Event geofenceEvent = analyzeGeofenceEvent(jtsLine, jtsPolygon);
```

---

[1] Note that the geofenceManager is created explicitly for this test, while this is automated in the handler route by Spring IoC.

[2] Map jargon for the line between two nodes, i.e. coordinates

The method analyzes whether the vehicle is entering or leaving the geofence and creates an `Event` entity with a type according to Traccar constants[3]. If the method returns an Event (`if (geofenceEvent != null)`) we can set the entity relations and finally attach the event to the device that will be persisted at the end of the handler route:

```
geofenceEvent.setPosition(lastPosition);
geofenceEvent.setServertime(new Date());        // creation time
geofenceEvent.setDevice(dbDevice);
geofenceEvent.setGeofence(geofence);

Set<Event> dbDeviceEvents = dbDevice.getEvents();
dbDeviceEvents.add(geofenceEvent);
dbDevice.setEvents(dbDeviceEvents);
```

Now the geofence manager is done with its analysis and the events have been attached to the `dbDevice`. This device can be picked up by the next manager in the route and the events will be persisted at the end of the route – possibly with more events from other managers and with the incoming events and positions from the `inDevice`.

17.5.2   Route Modification

We have just created geofence entry and geofence exit events in the same manner the Traccar system does. The event is attached to the position right *after* the event has occurred. With JTS it is easy to determine the exact coordinate of the intersection and calculate the (mathematically) exact time.

```
from                 geofence | intersection        to      positions
X----------------------X--------------------X
|<--      distanceOne    -->|<-- distanceTwo -->|         distances
t0                          ? t1 ?                 t2      timestamps
```

By calculating distances one and two you get the ratio, apply it to get the timestamp of the intersection and can finally create the exact position of the event!

Anyway we will not create this additional position. The reason concerns an unmentioned guideline for the complete tracking system. The vital question is whether you want to persist only the exact positions submitted by the tracker message or whether you allow modifications. Classical tracking systems usually do not modify original tracks while SDC systems aggregate tracks from many different vehicles to represent the traffic situation. These systems provide many utilities to optimize tracks by snapping coordinates to the road geometry, removing positions without useful additional information etc.

---

[3]`TYPE_GEOFENCE_EXIT` and `TYPE_GEOFENCE_ENTER`

17.6   ENTERPRISE STANDARDS

At this point we have learned how to apply different Java Frameworks to compose a system software with standalone modules. We have gone through some basic guidelines on how to apply a framework correctly and covered some common pitfalls. We, being the system architects, have set up the core modules used in a GPS tracking system. We have created some seed components that can be used as a starting point for a system dedicated to specific business cases.

There is much more you can do with JPA, Spring and Camel and hopefully you can use the JEETS components to apply additional features for your needs or as an entry point to go deeper into each framework. If not, we should meet at `jeets.org` to discuss requirements and put together a nice bare bone system that can run 'out of the box'.

This section concludes the creation of JSE components and outlines how to combine them to one system by using only the introduced techniques.

Enterprise development can be spread over different departments and locations. When creating software for connected car technology you have to design each module according to enterprise standards based on existing enterprise modules. Approved software should be available in *one* central Maven repository.

We have learned that a database entity relation model provides the *materialized* system model. Tables define attributes and relations ensure that you can not insert a message without an associated registered device and an account that you can send your bills to. Database engines are real work horses and every new table or attribute should be modeled into the database – continuously as the system evolves.

17.6.1   Data Formats

Once you have added your objects to the ERM you can connect to the database and generate the ORM. Of course you might as well start with the ORM and generate the ERM in the DBMS. The ORM is a (partial) representation of the ERM with Java entities that must be used by every developer to standardize his component. After the (initial) ORM generation you can manually fine tune the relations to be ordered, lazily loaded, cascaded etc.

The next step is to provide one or more persistence units for different programming contexts. We have created the JEETS PU basically to persist new messages in the database. Usually a GTS database model can be subdivided into tracking, administration, reporting etc. You have more control, if you create a persistence unit for each of these blocks. Or you can strip down the Traccar system to administration, reporting etc. and use your own DCS to collect messages.

Coding begins with the preparation of system entities from the incoming messages – the device communication. If you are using commercial trackers

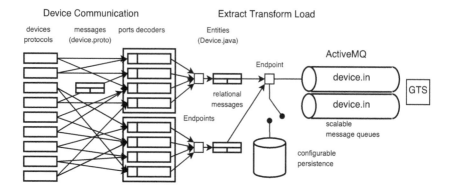

Figure 17.2 For high performance tracking device communication, extract, transform and load processes should be scalable by using different hardware for DCS, MQ, ETL and DB. The queue/s collecting incoming entities marks the entry to the actual GTS processing.

you usually can not define the tracker message format and need to program a message decoder, create the entity with the usual **new** constructor. With Google protobuffers you can derive relational messages to match the ORM and ERM and effectively insert the incoming data.

17.6.2 Message Transformation

With these fundamental data formats you can start coding the device communication and Extract Transform and Load the data to the system. With Spring and Camel you can make the DCS- and ETL modules completely independent and run each one from the command line. By introducing a message queue you can decouple the modules and even shut them down for maintenance in critical situations. This is very convenient for load balancing and scaling as you can run the modules on different machines. With modern deployment you could start additional modules at rush hours and shut them down when the traffic decreases to save machine power and costs.

After designing ERM, ORM, message formats, DCS and ETL the message queue becomes the entry point to the actual tracking system. On the other hand the device communication servers implement the core for tracking vehicles (or things)! After the system entities are sent to the message queue most components can be programmed against the database and a Web developer could access the database without any knowledge of tracking technology. For safety reasons the database access can be modeled with a REST API *without* CRUD operations. This way you can 'partition' resources by skill sets.

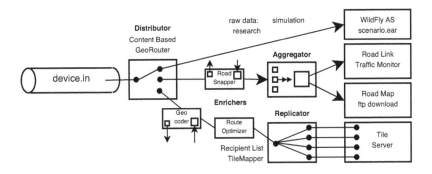

Figure 17.3 After incoming messages have been transformed to system entities these can be directed to various routes. With Camel you can distribute, enrich, replicate and aggregate the entity streams along the route to the targeted resource.

For a small business it might be sufficient to have a single MQ endpoint to serve all vehicles of all fleets. An enterprise creating a connected car platform has to deal with many concurrent messages and uses data distributers to send the messages according to geographical region (GeoRouter, tilemapper etc.), to dedicated OEM fleet software or, as we will see in the last part of the book, send the raw data directly to an application server for research. At system entry you should not clone messages to increase the load. Speaking in EIP patterns: Use the content based router and not the recipient list. The distributor serves as a semantic load balancer for the consecutive software modules.

If you want to add a traffic monitor to view the traffic you should look at the Aggregator EIP and how to implement it with Camel. Connected car technologies continuously create a 'live map' and vehicles can download their direct environment (tiles). To do this you can use Camel to place the map on an ftp server and each vehicle can use the map to enrich the onboard navigation with traffic information.

With DCSs we have standardized the relational messages to seamlessly work with every module. On the other hand you can create small, but very useful components to be applied on almost any Camel route. As an example you can build a 'road snapper' to snap the incoming positions to the road network. This can be achieved with the Java Topology Suite and the Digital Road map as an external resource. This tool could be downloaded from the Maven repository or it could be deployed as a service and a single map instance.

Persistence units are tiny components with large impact as they keep the system together as one. TileMapper and GeoRouter were created with existing entities, and it is vital to isolate PUs as individual jar files in order to distribute them to almost every module build on top of them.

When designing a system you should be aware of *where* you distribute, clone or modify the original information transferred with the message. While the georouter does not modify information, the road snapper does. Besides checking plausibility a route optimizer analyzes a route to return an optimized route reduced to positions with state changes. After moving original positions along the trace direction and speed attributes have to be recalculated.

Each of these components *can* be placed on every route but with respect to performance it should be carefully evaluated and not applied multiple times. If a Camel route does not need a geocoder don't place it there and save machine power.

Another pattern we have seen is the enricher. Like demonstrated with the geocoder, enrichers can easily be added to exisiting routes. As we have seen the Google geocoder produces a lot of data and the system designer should decide which route is suited for geocoding. For small companies an external geocoder is a cost factor and it makes sense to apply geocoding only *when* the information is needed for a detailed trip report, map display etc. Coding every message with a long address string has significant impact on the total database size.

17.6.3 GTS Components

We have thoroughly looked at the relation of emerging connected car technologies to 'classical' tracking systems. Now we will finalize the creation of JSE components and resume the classical GTS architecture. As a matter of fact the Spring framework (especially Spring Boot) is moving JEE standards to JSE components that can basically replace any application server environment. Individual components are easier to control by an operator and easier to monitor the OS processes individually.

Cloud computing is not just another word for a server; it describes the process of disassembling enterprise applications to individual modules and separating business customers by providing a complete GTS to each one of them – potentially combined with common resources like a database or message broker. While the Java coding of system logic does not change very much, the deployment process requires more programming logic. With an automated deployment process you don't need to touch every customer module for configuration etc. The module is created, debugged, and maintained by Java programmers and then it is tested and distributed to any number of servers by the operator.

Let's have another final look at our production system Traccar GTS which has perfectly served our purposes and complemented our JEETS components at development time.

17.6.3.1 Server Manager

We have identified Traccar's `ServerManager` as the Controller for all DC servers. Let's say you wanted to create a stand alone version of Traccar's `ServerManager` that you can use to supply system entities from different protocols. In an enterprise environment you could modify Traccar's `pom.xml` file to create a sub project to create a `traccar.protocols.jar` which includes the `default.xml` file with predefined protocols and ports[4].

Then you create a `ServerManager` project with a Camel or Spring startup that takes care of the bootstrap configuration for all DC servers, i.e. protocol and port pairs. The output of the `ServerManager` should be defined by a small number of system entities: the protobuffer protocols deliver ORMs to their endpoints while the Traccar protocols create position objects from different protocol types and formats.

17.6.3.2 Data Manager

Traccar's `DataManager` can be replicated with the JPA specification, the PU and the `EntityManager`. We have created the handler route to demonstrate entity handling along the route of different managers. In the case of JPA every `EntityManager` can be provided by an `EntityManagerFactory` for which Camel provides implementations of many (emerging) database technologies[5]:

Apache Cassandra database
for scalability and high availability without compromising performance.

Apache Couch database
Seamless multi-master sync, that scales from big data to Mobile, with an Intuitive HTTP/JSON API and designed for Reliability.

Dropbox
treat Dropbox remote folders as a producer or consumer of messages.

Gora
NoSQL databases using the Apache Gora framework.

Hadoop
For distributed storage and processing of big data datasets.

MyBatis
Performs a query, poll, insert, update or delete in a relational database

PGEvent
Allows for producing/consuming PostgreSQL events related to listen/notify commands added since PostgreSQL 8.3

---

[4]you should update each protocol from Netty3 to Netty4
[5]The main requirement is access via JDBC.

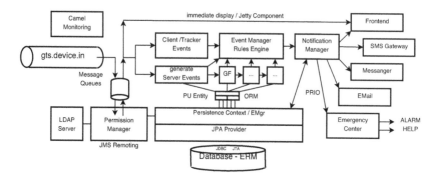

Figure 17.4 With Camel integration you can compose a complete GPS tracking system with Java SE standalone modules.

Couchbase
Working with Couchbase NoSQL document database.

... to list a few.

From performance considerations it makes a lot of sense to start entity processing with a database lookup with SQL or better JPQL for painless platform changes. Each GTS manager should be aware of the available information in the software created or retrieved from the predecessor and should navigate through the ORM to enrich it. Once the context is completed the last manager can persist every modification applied along the route and other handler routes can instantly synchronize the new info.

17.6.3.3 Permission Manager

In case of Traccar the permission management can be implied in every database query. If permission is denied the query returns an empty result set; if it is granted an exchange is sent on the GTS route. If you choose to use an application server you can use its security domain to create safe sessions. The login process can also check user and role permissions with SQL statements.

For large connected car systems this process is much more complex and messages must traverse the process of identification (who?), registration (listed?), authentification (digital identity) before it is finally authorized ('come in') to enter the system.

To simplify admission sequences Figure 17.4 shows an abstracted permission manager. This manager might deal with internal enterprise LDAP servers or require to use security domains etc. With 'hiding middleware' we have learned how to create independent Java software and how to add a Camel endpoint. No matter how complex the admission might be, it must return Yes or No as the result.

From the figure we can tell that the permission manager acts as the door keeper by directing each message to the permission queue. How does this work? The technique to apply here is called 'JMS remoting' and is supported by Camel and Spring (Remoting) although it is actually provided by message queuing.

Tutorial on Spring Remoting with JMS [48]

"This tutorial is a simple example that demonstrates more the fact of how well Camel is seamlessly integrated with Spring to leverage the best of both worlds. This sample is a client server solution using JMS messaging as the transport.

The server is a JMS message broker that routes incoming messages to a business service that does computations on the received message and returns a response."

With respect to our permission management we can create an MQ endpoint to route messages to role-, rule-, group permissions with authentication and return the message that can be forwarded to the tracking system.

17.7 USER INTERFACE

Once the message has entered the GTS it is time to check the message content and derive actions to be taken – ASAP. When we speak of tracking we want to know where the vehicle is *now* and in a transport company you want to see it on the map. So far we haven't dealt with a Webfrontend at all[6]. If you have a closer look at the Traccar architecture you will find that the core Traccar GTS (that we have imported into Eclipse) doesn't include a frontend either. Traccar provides a nice REST API to talk to the server and to build your own frontend with.

---

[6]SDC scenarios usually (as of today) don't provide graphical user interfaces to consumers for many reasons.

Inside Traccar you will find the class `AsyncSocket` which forwards entities with the method

```
private void sendData(Map<String, Collection<?>> data) {
   if (!data.isEmpty() && isConnected()) {
      try {
         getRemote().sendString(Context.getObjectMapper()
                     .writeValueAsString(data), null);
      } catch (JsonProcessingException e) {
         Log.warning(e);
      }
   }
}
```

For Camel developers this method demonstrates how to connect to the webserver, transform the entity to a JSON object and send it. So let's check the Camel home to find:

Camel Jetty component [49]

> The jetty component provides HTTP-based endpoints for consuming and producing HTTP requests. That is, the Jetty component behaves as a simple Web server. Jetty can also be used as an HTTP client which mean you can also use it with Camel as a producer.

In other words you can send messages to an existing Traccar frontend!

17.8 EVENTMANAGER AND RULESENGINE

Much more important than seeing the vehicle position on a map is the event management. If we take a look at Traccar's handlers as listed on page 124 we can locate a number of event handlers:

```
     overspeedEventHandler = new     OverspeedEventHandler();
       motionEventHandler = new       MotionEventHandler();
     geofenceEventHandler = new     GeofenceEventHandler();
        alertEventHandler = new        AlertEventHandler();
     ignitionEventHandler = new     IgnitionEventHandler();
  maintenanceEventHandler = new MaintenanceEventHandler();
```

First we need to distinguish tracker- and server events. If a person tracker sends an `ALARM` event the system should forward it immediately in order to assist the person fast. As this exceeds the core of tracking systems we need to create a high priority route to forward these events to a monitor center for telecare. We have created our own `GeofenceManager` to create `ENTER` and `EXIT` server events as an example of an event evaluated on the server side.

In Figure 17.4 every client and server event is forwarded to *the*

`EventManager` as a place holder for event handling. In GTS literature you can also find the term `RulesManager` to indicate complex algorithms. For example you might want to evaluate sequences of events to derive an event trigger. A traveling salesman can be instructed to visit customers in a predefined order. In technical terms we can (dynamically) create one geofence for each customer with a rule to verify the visiting order. This is not as straightforward as one single event but requires more programming logic including some bookkeeper object to collect relevant events...

Depending on your GTS market for people, vehicles or things you can often face complex rules defined by your customers and the Rules engine is the place to implement them.

17.9 NOTIFICATIONMANAGER

And finally, what's the use of a tracking system, if no one receives the events? The forwarding of messages *from* the GTS is implemented in the `NotificationManager`. Again, you can start by checking for Camel components like:

Google Drive and -Mail and GMail

camel-mail
Receiving email using POP3 and JavaMail
Sending email using SMTP and JavaMail

SMPP
To send and receive SMS using Short Messaging Service Center using the JSMPP library

Facebook

camel-twitter

17.10 PUTTING IT ALL TOGETHER

We have now gone through technologies to create tracking components and we have looked at the major Traccar components and evaluated ways to migrate them to a new standalone component to be used in various systems like GTS and SDC.

We have not covered the graphical user interface and we will not cover the `ReportManager`. Creating reports is not really related to GTS technologies and can easily be created with an independent Web frontend – against the database. Automation is definitely a progressive strategy to control the CPU power and database performance from interactive report generation during office hours. Reports can be automated by triggering weekly and monthly reporting, Java developers can create PDF documents with the PDFbox and send them by mail.

Apache PDFBox - A Java PDF Library [61]

> "The Apache PDFBox library is an open source Java tool for working with PDF documents. This project allows creation of new PDF documents, manipulation of existing documents and the ability to extract content from documents. Apache PDFBox also includes several command-line utilities."

This book was written to introduce tiny while powerful JEE components and frameworks and show how to combine and use them in an JSE environment[7]. The JEETS repository is not an 'out of the box' application like Traccar. Yet JEETS components can be used as a starting point for a system with a dedicated purpose.

Now you can take a scratch piece of paper, draw you routes and then look into Netty, Jetty, Spring and Camel to combine them in your implementation. The tutorial on JMS Remoting is based on the skills we have established. Find out how to prioritize routes in Camel and how to monitor distributed Camel components[8].

Another topic to cover are Maven integration tests. With them you can run different Maven builds with plain JSE, with a real database, with an MQ broker and so on. The main idea of integration testing is to test the interaction of different modules and in the end the complete test can setup trackers or players to send messages and then trace the message processing all the way to the system output.

We have almost reached the end of the book and have gone through many technologies to track anything with individual components controlling external resources. In the last part of the book we will build a bridge to an JEE application server and open the door to a complete JEE environment. Before heading on you can use the chance to prepare your environment for hands on experience with WildFly as described in Appendix E.1.

---

[7]...and in a JEE environment in the next part.
[8]We will introduce hawt.io in section 18.4.4

Please prepare your environment to send messages
to a message queue from more than one device.

In order to do this you can prepare command lines
for one or more JeeTS trackers, players and/or clients.

You can set up a GTFS datasource to simulate public transportation.
With the GTFS set from Hamburg you can create factories
for any number of subways, buses, ferries etc.

Then you can setup a DCS module to receive these messages
and forward them to a message queue
(Persisting with ETL etc. is *not* recommended at development time.)

In another step you can apply a georouter to distribute messages
for your selected GTFS area to another message queue.
That will be your entry point to use for the application server.

VI

Enterprise Applications

The development of information technology from the first unix system in the late sixties to DevOps has gone through various paradigm changes. Software developers have always striven for a seamless development cycle from the very beginning and with continuous development this claim has been fulfilled – after some laborious decades! Software development combined with the deployment on hardware is exponentially accelerating software innovations and drastically reducing time to market. Users are not bothered with complicated update procedures.

For prototyping its always a good idea to create one or more pieces of software to accept input and provide (hard coded) output – before implementing details. We have gone through the major modules of a tracking system and created stand alone prototypes for them. For new team members this approach offers transparency of the actual information processing and allows the project manager to direct work and (temporarily) employ experts by domain skills.

On the other hand the JEE specifications are essential to guarantee compatibility in complex software systems. Specifications are invaluable (for enterprises), but what about their implementations? A JEE application server is only certified, if it provides *all* JEE specification implementations. From a high level view we can see that you can setup Spring with *any* JEE specified component *or* you can create one enterprise application (business logic) and deploy it into a complete preconfigured environment.

The approaches of Spring JSE components and enterprise applications (ear) in application servers (AS) are complementary to each other and have a large impact on development and deployment. While the ear file is deployed to the AS in one piece the Spring components can be deployed individually. With the deployment of many Spring component instances the configuration (and work) moves from Dev- to -Ops[9]. We have seen how to apply message queues as endpoints between modules, but at least the common MQ names and the persistence unit have to be configured for every JSE component... preferably with a build tool.

Another good reason to modularize a system is the fact that computer speed will not significantly increase any more. Moores Law has correctly predicted CPU development for three decades but this development has reached physical limits. Today computer speed is increased with the number of CPU cores called horizontal scaling. Instead of using complex AS or DB clusters you can simply instantiate your components (DCS, ETL etc.) multiple times on different machines.

Visualization is a great support for operators as they can prepare preconfigured environments with resource installations (AS, DB, MQ etc.) and clone them as many times as needed. In the end scaling by Cloud, VM, Docker, AS etc. should be done with respect to the *hardware resources*. What's the use of two databases in two VMs, if both share the same CPUs? Cloud computing is about distributing electric power over many small software modules. On the

---

[9]tied together by continuous testing, integration and deployment.

other hand, the computer industry can only grow with sufficient energy and is competing with classical consumers like cars – and people. Just think of the billions of smartphones as standalone high tech computers, always up and running, demanding a complete load every day!

The -Ops part of DevOps was also driven by administrative and hardware developments. During the nineties the hibernation of complete operating systems with installed and running software turned out to be a great challenge – while it was a barely noticed convenience for PC users. After hibernation the step to virtual machines was relatively small and modern CPUs come with a virtualization mode in the BIOS. VMs turned out to be very convenient. On the other hand a hibernated VM is only a binary blob and it shouldn't be misused for (incremental) backups as you can not access its frozen content. Databases should backup data on a regular basis and to enable developers to search for datasets in the archives.

Docker can be perceived as the next step to more transparent deployment, but we will not go into the details here. The point is that the traditional operating system is abstracted and reduced to a 'container', which provides the software environment. From Java developers' perspective the containers are slowly being separated from and moving out of the application server.

This book is about software development and how to separate business logic (Dev) from deployment (Ops). A software developer is not in charge of leveraging the CPU power in production systems. Why should he investigate the number of (virtual) CPUs in his Dev VM, if he doesn't know what CPU slices will be available from the production hardware? He should rather provide some benchmark tests to provide hints for the operator. We have seen how Camel can be used to split and merge processing steps and how Spring creates objects via IoC to handle the load and adds Aspect-Oriented Programming (AOP) to complement Object-Oriented Programming (OOP).

The operator is in charge of the system behavior in general and modern cloud computing provides means to scale a system. SDC systems *require* this kind of hardware scaling and have challenged software development to make it happen! We all know the critical traffic spots from our daily trips and we know the rush hours when to avoid traffic. The same applies to cloud computing: During rush hours the system can automatically launch additional instances of any module.

A modern tracking system should be able to raise CPU power on demand and lower it to control the hosting costs – electricity in the end. Or more consequently: a modern tracking software can be set up for every (major) customer! It's only a matter of creating instances. You can launch many DC servers on the input side while sharing one database with the system model and one MQ broker for all customers' GTSs.

CHAPTER 18

Spring to AppServer

CONTENTS

18.1   SPRING VS. JAVA EE

In the late 1990s Java started to adapt to technologies like JTA and introduced RMI etc. for large scale applications. These technologies called for a component framework for a safe and distributed multi user environment. The EJB 3.2 specification [69] with more than a thousand pages defines a component model and a container framework. Anyway, application server implementations differ and the developer is tempted to apply supporting, but proprietary APIs, which make it hard to change AS providers, i.e. Weblogic to WildFly etc.

Choosing the development and runtime environment for your system is a vital for your business modeling process. We have seen how to create our own dedicated JEE components with the most important specification JPA to standardize all interactions with the system model. The persistence unit *carries* the system model to any chosen environment. Historically the JPA entities were part of the EJB framework until they were migrated to the JPA framework due to their importance as (light weight!) entities for a complete system. The application server represents the runtime environment for many different framework components – to spare the developer from setup and configuration work.

Spring Boot is a new concept to compose and run complete applications without an AS. These concepts have been applied much earlier in other languages – due to the lack of an AS. Spring Boot provides pre-configured technologies to reduce configuration – like an AS, only more dedicated to the deployed application. With Spring Boot you can quickly get your application running with very little business code and out of the box frameworks.

A very important aspect of modern development is the need for innovation and adaptation. With the Spring framework you can focus on functionalities required for new approaches and picked up by the community. Therefore Spring can be seen as a lively Sandbox for experiments with new challenges. When the new approaches are accepted, become successful and productive, they will probably be standardized in JEE sooner or later. Some say that CDI was invented with Spring IoC.

In the end you should prefer Spring for flexibility of design, while you should use an JEE AS to ensure official standards. If you have legacy JEE applications don't worry, JEE is alive and application servers are extremely convenient (for prototyping). Spring is one technique to extract code from an AS and run it stand alone. WildFly Swarm is a technique somewhere in between and closer to AS and uber jars is closer to Spring etc.

Confused? If you look back at the JEETS components DCS, ETL etc. you can easily identify the business code for tracking. This code should be developed in a neutral style to interact with its environment, be it application server or Spring Boot. If you carefully study the differences you will get a feel of how to create core code like classes, projects, JEETS components etc. and embed them in either Spring or as part of an enterprise application (ear).

After all we are Camel developers and are able to connect different components running in different environments. We'll see how to combine our JEETS components with an AS environment soon.

18.2   APPLICATION SERVER

The initial intention of creating application servers was to provide all JEE specified implementations and interfaces in a single environment. The AS running on a JVM actually is a pure *Java server* environment with Java interfaces to the rest of the IT world!

For JEE developers the application server is the environment for various containers that can communicate with external resources in the Java language. The AS can be perceived as a software environment providing and combining subsystems like JPA, CDI, JSF and the heavy weight EJB with JTA, security aspects, etc. Java has pioneered deploying software in an (J)VM, which can completely hibernate an AS (for seconds) to collect garbage, execute cleaning jobs etc. This *Java server* is very convenient for development and research, but...

Which JEE specifications will we really need in production?

We saw how to apply the important JPA, JTA, JMS specs etc. in JSE. On the other hand we know today that specifications like JSP and JSF have not been accepted by frontend developers. You will have a hard time finding a Web agency to create a modern single page application (SPA) with JSF (application server) or even JSP (Servlet engine). Traccar is showing a better way by supplying a REST API for Web development[1]. On the other hand JSF is the fast lane for *Java* developers to visualize information in the browser without knowing JavaScript frameworks. Therefore it can be a good choice for *internal* monitoring of the complete system throughput.

So why use an application server?

In the context of SDC we have to deal with big data amounts and create different pipes and routes to direct the message flow. The SDC system has to be tuned for best performance and behaves erratically, if the processing can not keep up and catch up with the incoming stream. We can direct all (virtual) messages from public vehicles to one 'simulation environment'. We have already created an environment to replay one or more tracks that we have recorded (with Traccar). As a result we can get a scenario of different vehicles moving in an explicitly defined bound box from a single intersection to a city, country or whatever. Figure 17.3 on page 209 provides an overview of what we're up to. After applying DCS and/or ETL to receive messages we can forward system entities to an AS environment.

Traccar is a monolithic JSE software from device communication to browser, map frontend and notifications – completely focused on core tracking technology. When you establish a GTS business the customers begin to develop new ideas for new devices. Drivers should interact via the touch screen of their navigation system, Employees should be tracked via smartphones and raise the need for identification and authorization with driver cards, etc.

Once the tracking system is feature complete the customer urges *integrating* existing systems like ERPs etc.

> For logistic companies it can be extremely helpful to use their vehicle- and driver management systems with live locations.

> For tracking people with Ambient Assisted Living (AAL) the system has to be reliably connected to an emergency response center.

> For person transport agencies it is critical to synchronize vehicles, drivers and passengers with the booking system.

> ...to name a few scenarios.

---

[1] www.traccar.org/traccar-api

As a generalization we will look at WildFly as another existing system to manage goods, vehicles, drivers combined with secured billing systems, logistics or whatever business you may be in. And we will focus on the message route from the port as the hardware endpoint via DCS or ETL directly into an enterprise application deployed in one EAR file.

In the SDC context the application server environment can also be used to simulate software that is actually embedded in the car's hardware. By feeding the existing system with live vehicle information surrounding vehicles can be located and the tracking information can be fed to the in-car software and simulate different sensing technologies to perceive the environment. You can even simulate a complete car and virtually hit the breaks when a sensor receives signals from approaching vehicles. Beyond that you can simulate a server software to observe critical traffic spots and trigger system scaling according to traffic.

Please read Appendix E.1 to install WildFly.

18.3   CAMEL AND WILDFLY

So far we have created some JEE specified JEETS components to be controlled (start, monitor, stop ..) via command line on the OS level. This final part about JEE applications will demonstrate how to create JEETS components for the application server. For the knowledge domain of GPS tracking it should not matter, which runtime environment or container you chose. We will see that Camel can easily integrate these environments to one *bus system* for *routing system entities*. You can develop your components, i.e. Maven projects, and deploy them in standalone Java SE, CDI containers or JEE application servers with EJB containers. Then you should be good to go for your own – existing – system.

Why is WildFly not listed as an external Camel resource?

From the previous pages we can deduce that OS, VM, Docker and the AS 'Java computer' are simply different software *environments* to run business code in. Traditionally the application server was developed to abstract the outside world with Java APIs. When one application server is overloaded with information streaming and processing, scaling comes into play. Yet the development of AS clustering (DB replication etc.) has turned out to make things more complex. Therefore the trend was reversed to use the application server for a carefree development environment and strip it down to the necessary components at deployment time – see WildFly Swarm:

WildFly Swarm - Right Size Your Services [55]

> "WildFly Swarm offers an innovative approach to packaging and running Java EE applications by packaging them with just enough of the server runtime to "java -jar" your application. It's MicroProfile compatible, too."

The relevant environment for business developers is the container!
Containers are managed and provided by operations personnel, while a classical JEE AS was not created to instantiate deployments more than one time. Modern systems scale 'horizontally' by adding more machine power and creating additional container instances. With Camel things become easier to grasp by simply providing endpoints and herewith ignoring, i.e. hiding the environment! Any component can be deployed in Spring, Tomcat or WildFly and will be addressed only via Camel endpoint URIs from enterprise integration patterns.

How can we integrate Camel in WildFly?

With a critical analysis or from implementation experiences you can find out that application servers provide, but do not enforce JEE specifications. Depending on your constraints you *can* add Camel components to your AS deployments without touching the WildFly installation and configuration. The AS allows you to bootstrap all of your Camel components, even with Spring XML and other dependencies. This can be helpful for migrations, but leads to errors which are hard to locate and the artifacts (jar, war, ear) become bigger with overlapping contexts[2]. If you decide to use Camel for your complete system in the first place then you should add the WildFly-Camel subsystem to your WildFly server – it's simple.

18.3.1   The WildFly-Camel Subsystem

Simply speaking the WildFly-Camel subsystem integrates the complete Camel framework with your 'WildFly Java computer' to make its interior, other subsystems, addressable for an external system bus. In this constellation you can deploy routes as part of a JEE application or its components as part of a route with access to any Camel component API. Behind the scenes the Subsystem itself integrates tightly with JEE and routes can be deployed as part of enterprise applications (ear) while its components can access any Camel component API. The subsystem enriches WildFly with Camel features without additional configuration or deployment. If you look at the admin console of WildFly (Configuration > Subsystems) you can see that the AS orchestrates many subsystems. EJB is the core subsystem closely mingled together with the security, transaction subsystems, etc.

With WildFly-Camel we can also stick to our main Camel version in the

---

[2]This problem has been addressed with Java 1.9

repository to maintain versioning in one place tested and provided by the Camel Riders. The decision to apply WildFly-Camel has impact on the system *architecture* as you can combine any component inside AS with Camel routing. Therefore Camel is often referred to as ESB-lite.

Please read Appendix E.2 to install WildFly-Camel.

18.4   JAVA MONITORING

Up to this point we have invested a lot of research to create single JEETS components. This book does not compose a complete GTS, but rather describes the migration process from traditional monolithic to distributed software. By splitting a system into individual components the operators responsibility rises and he will need to establish monitoring tools and constraints to observe the system and trigger some kind of scaling.

18.4.1   WildFly Console

In WildFly's admin console you can find different monitors to check the current status of the JVM.

18.4.2   Mission Control

If you go to the root directory of your Java installation you can find the Oracle Java Mission Control in the `bin` folder, i.e. `.../jdk1.8.0_60/bin/jmc.exe`

Oracle Java Mission Control

"Oracle Java Mission Control is a tool suite for managing, monitoring, profiling, and troubleshooting your Java applications. When first installed, Java Mission Control consists of the JMX Console and the Java Flight Recorder. More plug-ins can easily be installed from within Mission Control."

If you launch Mission Control you can click on the running WildFly instance and inspect Heap, CPU etc. and you can even send some diagnostic commands. This can be helpful at development time, but is very cumbersome for the operator. The operator would have to navigate into the OS environment (does he have the rights?) for every single running JVM. Haven't we created a distributed environment for *more and better control* of the system behavior?

18.4.3 Java Management Extensions

Java Management Extensions [70]

> "Java Management Extensions (JMX) is a Java technology that supplies tools for managing and monitoring applications, system objects, devices (such as printers) and service-oriented networks. Those resources are represented by objects called MBeans (for Managed Bean). In the API, classes can be dynamically loaded and instantiated. Managing and monitoring applications can be designed and developed using the Java Dynamic Management Kit."

We will not go into the details, but you can re/visit the Spring remoting tutorial suggested on page 213 to find the section 'Using Camel JMX' [71]. If you look at the screenshot in this section you can identify the same MBeans you have already found in the Mission Control display. Note that is up to the developer to provide and enable the JMXAgent in order to introspect the software at runtime.

18.4.4 Hawt.io

The above list of monitoring tools can be helpful in special situations. But we have to consider the movement of responsibilities between Dev and Ops and it would be much easier to provide a tool to connect to any JVM and inspect the ongoing processing live. And that is exactly what Hawt [72] is about.

Hawt - A Modular Web Console for Managing your Java Stuff

> "hawtio has lots of plugins such as: a git based Dashboard and Wiki, logs, health, JMX, OSGi, Apache ActiveMQ, Apache Camel, Apache OpenEJB, Apache Tomcat, Jetty, JBoss and Fuse Fabric
> You can dynamically extend hawtio with your own plugins or automatically discover plugins inside the JVM."

You probably have noticed that WildFly-Camel is installed with Hawtio. Check WildFly's startup log for the line

```
INFO  [org.wildfly.extension.camel] ...
        Add Camel endpoint: 127.0.0.1:8080/hawtio
```

At `localhost:8080/hawtio/` you can login with the WildFly application user you have created earlier.

Please read Appendix E.2.2 to log into Hawtio,
inspect WildFly and connect to ActiveMQ.

What we see in the console is a preconfigured monitoring tool installed on and coupled to the application server with a preconfigured and coded Camel subsystem to monitor every Camel route and component! The console can be used by everyone who has access to the WildFly URL. For your own studies you can find a complete test with Hawt and WildFly at

`github.com/wildfly-extras/wildfly-camel/blob/master/docs/Hawtio.md`

Note that you can find the WildFly-Camel subsystem in the WildFly frontend, but as of now we can not inspect any Camel components simply because we haven't deployed any. Let's do so.

18.5   WILDFLY-CAMEL QUICKSTART

We have basically gone through the 'WildFly Camel User Guide' getting started section to prepare our environment. You will also find two Maven archetypes:

WildFly Camel User Guide [54]

"To get started with writing Camel JEE applications, there are two Maven archetypes that can generate either a Camel Spring XML or Camel CDI application."

If you have installed everything it is time to validate the WildFly-Camel installation by deploying a war to WildFly and visualize the Camel route with Hawt. You can direct your command line to some (temporary) folder and create a pure WildFly-Camel project with the command line

```
mvn archetype:generate
   -DarchetypeGroupId=org.wildfly.camel.archetypes \
   -DarchetypeArtifactId=wildfly-camel-archetype-cdi \
   -DarchetypeVersion=5.1.0 \
   -DgroupId=com.mycompany \
   -DartifactId=my-camel-cdi-application
```

Note that Maven even provides a `README.md` file with instructions to compile, test (see next section), install and deploy the project artifact. You should manage to execute the following steps on your own:

```
mvn clean install
mvn clean package wildfly:deploy
access the application via browser
(mvn wildfly:undeploy)
```

If you access the application with

```
localhost:8080/my-camel-cdi-application?name=Kermit
```

you send your first message on the route and ensure that you have successfully prepared WildFly to work with Camel and you can finally see the Camel route. The Hawt frontend automatically adds the Camel tab and you can inspect the route, the endpoints and even watch the messages traversing the component live.

18.6   DEVELOPMENT WITH ARQUILLIAN

In the previous section we skipped the Maven step to test the application. Testing inside an AS environment has been a real pain for years until the JBoss community came up with the Arquillian test framework [56]. The idea of Arquillian is to deploy an isolated test project archive to an AS and ignore all of the 'official' deployments.

The application you have just generated and deployed uses a @WebServlet, i.e. HttpServiceServlet as browser endpoint. The test class listing reveals how to create a test.jar and deploy it. Then Arquillian takes over and sends a message with Camel's ProducerTemplate to the route defined in jboss-camel-context.xml. Note that Arquillian requires an AS home variable to find the AS. You can choose from the Maven profiles 'remote' for a running instance or 'managed' to instruct Arquillian to also start and stop the (test) server.

If you are aware of test driven development then Arquillian is a powerful tool in which to start development in. This makes the developer's life much easier as he can focus on the one implementation he is working on instead of getting confused in a complex AS setting.

18.7   WILDFLY-CAMEL SPRING

With the CDI Camel WildFly example we have verified the functionality of our 'AS integration suite' that is ready for GTS development. You should do the same with the second Maven artifact to find out how Spring works in the JEE environment. Generate the WildFly-Camel-Spring project with

```
mvn archetype:generate
    -DarchetypeGroupId=org.wildfly.camel.archetypes \
    -DarchetypeArtifactId=wildfly-camel-archetype-spring \
    -DarchetypeVersion=5.1.0 \
    -DgroupId=com.mycompany \
    -DartifactId=my-camel-spring-application
```

and inspect the slight differences from the previous CDI application. The Spring project uses the `jboss-camel-context.xml` file instead of a CamelRoute class and the Servlet has an additional line

```
@Resource(name = "java:jboss/camel/context/spring-context")
```

to access a `@Resource` via JNDI lookup.

With these two projects you could basically go back to the beginning of part V 'Middleware' and migrate the JEETS components from an OS to an AS environment. We will follow our mission defined with Figure 17.3 on page 209 to process filtered messages inside an AS environment. In order to do this we need to prepare another constellation – WildFly-Camel interaction directly with ActiveMQ.

18.8   WILDFLY-CAMEL ACTIVEMQ

We have deployed a WildFly-Camel CDI and a WildFly-Camel Spring project and now we will add an WildFly-Camel ActiveMQ project. The difference is the external broker resource, which needs to be configured separately. Again we can apply modern software development by going through another WildFly-Camel example provided as open source at [58].

You can download the complete examples repository and go through the `camel-activemq` hands on example to be prepared for the JEETS implementation.

18.8.1   ActiveMQ Resource Adapter

With the WildFly-Camel installation we have added Camel's ActiveMQ integration by the `activemq` [51] component. We have already used this component for JSE implementations with MQ endpoints. Before we can use it with Wild-Fly we have to configure the endpoint URI for a named message queue.

Before DevOps it used to be the operator's responsibility to configure resources around the business software. The developer only had to know the endpoint URI and could be assured that it is provided in the system environment. For continuous integration the repository should be self contained to be able to install and configure a complete resource at build time. If you look at the WildFly examples you can locate the directory

```
github.com/wildfly-extras/wildfly-camel-examples/itests
```

This folder defines the integration tests for all examples and also takes care to setup WildFly, Camel and the ActiveMQ resource adapter. This way the installation, configuration and deployment is driven by the Maven build and is capable of populating a container – which has been created at runtime for horizontal scaling.

Please load the project
`.../jeets-server-jee/jeets-wildfly-activemq-adapter`
in your IDE and open the `README.md` file.

The project is subdivided for two purposes.

First the `activemq-rar.rar` is downloaded from an official Maven repository and with the correct version defined in the pom. This is convenient and only has to be executed once for every WildFly instance. On the other hand you don't have to mind the un-installation of the rar, but keep in mind that the `clean` step simply removes the complete `activemq-rar`.

Second, the project relevant installation is defined in the project file:

`.../src/main/resources/cli/configure-resource-adapter.cli`

This file takes care of configuring the ActiveMQ resource adapter for a device queue with the physical name `device.in`[3], the queue required for the JEETS `jeets-dcs-amq` project.

Please launch the deployment with the Maven profile with

```
mvn clean -Pdeploy-amq-rar
```

and you will remove the `activemq-rar` which can be counter checked in Wild-Fly's Web console. Restart WildFly and run

```
mvn install -Pdeploy-amq-rar
```

and counter check your WildFly installation file

```
wildfly-11.0.0./standalone/configuration/standalone-camel.xml
```

in section

```
<subsystem xmlns="urn:jboss:domain:resource-adapters:5.0">
```

to identify the `activemq-rar`, the `QueueConnectionFactory` with URL and the `ActiveMQQueue` named `device.in`. Restart WildFly to conclude the installation and open WildFly's Web Console

```
Configuration: Subsystems
  > Subsystem: Resource Adapters
    > Resource Adapter: amq-ra.rar
      > Configuration
      > Connection
      > Admin Objects
```

to verify configuration, Connection and admin objects.

---

[3]Of course you can override the name according to your needs.

18.8.2 JeeTS Repository Management

Managing the repository has become the pivot point for developers and operators. As demonstrated with the resource adapter Maven profiles *can* be prepared for configuration tasks of external resources. With different Maven profiles for different purposes you can orchestrate a complete system setup in a brand new environment.

Before dealing with WildFly AS we chose to provide different frameworks as Camel JSE components. A great advantage of using the Camel integration is the versioning of all subcomponents. Since the JEETS repo now includes WildFly, WildFly-Camel and the resource adapter the JEE subdirectory

```
.../repo.jeets/jeets-server-jee
```

has its own POM as the top level for WildFly details. Depending on your plan you should take some time to find the correct BOM matching your strategy. Basically you can choose them from `github.com/wildfly/boms`

```
wildfly-javaee7
wildfly-with-tools
wildfly-javaee7-with-tools
```

and then add the `wildfly-camel-bom` according to your Camel version for the rest of the repository.

Now you have completed the JEETS setup for Camel Routing with external resources. For a real world test you should launch ActiveMQ and WildFly manually and we can finally create an enterprise application with live tracking.

The JᴇᴇTS EAR

CONTENTS

This book has demonstrated how to implement Java frameworks outside of an AS environment. It might have reflected the chronological development of software methodologies better to have started with a "JEE GTS" and then migrate, i.e. un-integrate the components to a plain JVM. The point is that most Java frameworks can be used in JSE *and* JEE. The choice should be related to requirements, manageability, granularity and many other environmental constraints – according to your design (constraints). The difference (only!) lies in the runtime environment (or context). In an application server the Java frameworks are provided as well coordinated subsystems with security etc.

This final part of the book will demonstrate how to create an enterprise tracking application in a single EAR file (i.e. `gts.ear`) and use the preinstalled subsystems in the AS. We have introduced tracking technologies by implementing JᴇᴇTS components and we have applied Camel integration to form a system with them.

These approaches are sufficient to create any customized GTS!

By creating one single EAR file we now want to explore and reflect the

JEE environment inside an preconfigured application server. In the end of this chapter you can create the EAR from the JEETS repository to implement your own JEE tracking scenario or system. The EAR is supplied as a seed component to *begin* tracking development. Once you get familiar with this environment you can choose between a JSE- and JEE environment or a mixture of them.

19.1 JEE ENTERPRISE ARCHIVE

EJBs are complex constructs (under the hood) [69] that can be deployed in a single file to an AS with the EJB framework configured to access various preinstalled subsystems. EJBs are easy to create, but when deploying a number of EJBs their management, services and interaction become more complex. An Enterprise ARchive is a construct to package a number of cooperating modules in one package.

The AS 'knows' exactly how to deploy the EAR according to XML deployment descriptors etc., the order of installing and running the components in the EJB container with given dependencies. The deployment process already implies a lot of validation itself. The EAR structure matches the AS structure and an EAR deployment to AS is comparable to that of a Docker component to its container.

Another comparison are the different deployments of an JSE archive with or without dependencies. If you look at the packages of the `jeets-tracker` target folder you'll note that the actual business code has a size around 20 kByte, while the `jar-with-dependencies` has about 20 MByte! The latter can be compared to an AS with installed dependencies.

In short: For a developer the EAR file represents the AS environment, which has been setup by an operator with external resources, i.e. database access via JPA with DB Pooling, secure multi user (Role) Login for customers and employees (LDAP etc.) ... developers can define EJB connectors declaratively with Java annotations or alternatively with XML descriptors.

With the JEETS EAR we want to supply some major technologies for your convenience and provide hints to add your own. The configuration of these technologies should ease your entry to an AS. Again, one of the main reasons to implement tracking for an AS, is the existence of other major systems in a company. Therefore we will not create a complete application, but you will see how to add live tracking information to your existing EJBs – how to integrate your system with tracking information. Let's see how we can abstract the environment for this general purpose.

Without describing every step we will now outline the rapid development of a subdivided ear file. You can then choose to use the JEETS EAR or create your own from scratch by following the instructions.

Please load the project .../`jeets-server-jee/jeets-jee-app` in your IDE. This project includes the subprojects /`ear`, /`ejb` and /`web`, which should be loaded with the main /`app` project.

Make sure to run `mvn install` on the main project to supply all dependencies via Maven (local) repository. Afterwards you can selectively work exclusively on a single subproject in your IDE.

With `mvn wildfly:deploy` (and `..:undeploy`) you can directly un/deploy to your WildFly instance outside of the IDE. Make sure WildFly and ActiveMQ are running with their respective frontends or Hawt in the browser.

19.1.1 ejb-in-ear

Step One

Install the following WildFly Quickstart[1]

```
github.com/wildfly/quickstart/tree/master/ejb-in-ear
```

What do we get?

We get an EAR project with a WEB subproject accessing the GreeterEJB in the EJB subproject via JSF and CDI subsystems. JSF pages access EJBs via managed (and scoped) beans.

Not bad for a start.

The EJB project contains the EJB code and can be built independently to produce its own JAR archive. This EJB jar should represent an existing business system which should be enabled to receive live tracking information.

The Greeter bean is the most basic EJB you can create:

```
/** A simple EJB without an interface. */
@Stateful
public class GreeterEJB {
```

For our application scenario we will create a main application bean with

```
package org.jeets.ear.ejb;
   :
@Singleton
@Startup
public class ApplicationBean /* implements MyApplication */ {
```

---

[1] Deployment of an EAR Containing a JSF WAR and EJB JAR

This EJB will be instantiated at deployment time should serve as the single entry of our tracking messages. From there you can dispatch messages to your components via Java (Messaging etc.), Camel or Spring. To achieve this we simply define the method which should addressed inside a Camel route.

```
public void processMessage(Device devMsg) {
```

This method receives all `Device` messages as system entities and from there on you can define the message flow through your scenario.

19.1.2   war-in-ear

The WildFly Quick Start `ejb-in-ear` also includes the `war`, i.e. `web` project. The `Greeter` is a simple managed CDI Bean

```
@Named("greeter")
@SessionScoped
public class Greeter implements Serializable {

    @EJB
    private GreeterEJB greeterEJB;

    public void setName(String name) {
        message = greeterEJB.sayHello(name);
    }

    public String getMessage() {
        return message;
    }
}
```

which injects the GreeterEJB and to invoke the method to say "hello". Next the CDI bean can be interactively addressed from a JSF. For example

```
<h:inputText value="#{name}" />
<h:commandButton value="Update" action="#{greeter.setName(name)}"/>
<h:outputLabel value="#{greeter.message}" />
```

...as you can see in the `greeter.xhtml`.
Accordingly we create a Controller CDI bean for our application:

```
@ManagedBean(name="appcontroller")
@ApplicationScoped
public class ApplicationController implements Serializable {

@EJB
private ApplicationBean appBean;
```

We are prepared to create a JSF frontend to monitor the application with the (JSF-) Expression Language (EL). We can simply add expressions starting with `#{appcontroller.<method|attribute> }` in a JSF.

Now we are well prepared to actually receive `Device` entities.

How can we achieve this?

19.1.3 amq-in-ear

In section 18.8.1 we have configured WildFly with an ActiveMQ resource adapter and a Camel endpoint to receive MQ messages. Nevertheless we haven't applied the project war

```
jeets-wildfly-activemq-adapter-x.y.war
```

yet. If you look at the `ActiveMQComponentProducer` class it is not much of a project. We have chosen this deployment structure to isolate the `java:/ActiveMQConnectionFactory` and to make the Camel endpoint between WildFly and ActiveMQ apparent.

In this respect JEE resources are similar to resources defined with an URI in a Camel route. The application can make use of a number of preconfigured resources with the `@Resource` annotation and connect with `@Produces`.

```
@Resource(mappedName = "java:/ActiveMQConnectionFactory")
private ConnectionFactory connectionFactory;
                        -----------------
@Produces
@Named("activemq")
public ActiveMQComponent createActiveMQComponent() {
    ActiveMQComponent activeMQComponent =
    ActiveMQComponent.activeMQComponent();
    activeMQComponent.setConnectionFactory(connectionFactory);
    return activeMQComponent;                -----------------
}
```

Recall the configuration of the message queue name in our `jeets-wildfly-activemq-adapter` project:

```
configure-resource-adapter.cli
jndi-name="java:/hvv.in"
```

With this pattern you can add any sources provided by your enterprise. In the project you can also find the *request scoped* resource definition to the JSF context:

```
@Produces
@RequestScoped
public FacesContext produceFacesContext() {
    return FacesContext.getCurrentInstance();
}
```

Step Two

Add the `jeets-wildfly-activemq-adapter` via EAR pom to the EAR lib folder. The lib folder can be perceived as a resource to all subcomponents of the EAR file. With the adapter we can receive messages from the ActiveMQ broker and create Camel endpoints handled by the WildFly Camel subsystem to feed them in our application.

For many different constellations you should visit the quickstarts at [58]. For example your existing applications might be working with JEE Message Driven Beans (MDB) to exchange messages with a broker. Then the example [59] should be of interest.

19.1.4   jpa-in-ear

Always keep in mind that the tiny `pu.jar` ties the components together and provides a *standard* for system wide compatibility. Therefore we need to prepare WildFly to provide JPA management for the `jeets-pu-traccar` and connectivity, i.e. data source to the Traccar database. Again it is important to remember that you could also use the Java entities of the persistence unit as plain POJOs without a database connection to avoid hardware latency, etc.

In order to use the PU in the AS environment the database credentials have to be installed there by operations. The changes for the PU project are minimal and only concern the `persistence.xml` file. The following properties...

```
<persistence-unit name="jeets-pu-traccar-jpa">
<properties>
  <property name="javax.persistence.jdbc.driver"   value=.. />
  <property name="javax.persistence.jdbc.url"      value=.. />
  <property name="javax.persistence.jdbc.user"     value=.. />
  <property name="javax.persistence.jdbc.password" value=.. />
        :              :
</properties>
```

...are *not* required for the persistence unit, since these properties have to configured in WildFly or you can pass the configuration with your deployment and it looks something like this

```
<datasource jta="true" jndi-name="java:/PostgresTraccar"
    pool-name="PostgresDS" enabled="true" use-ccm="false">
  <connection-url>jdbc:postgresql://localhost:5432/traccar3.14
                         </connection-url>
  <driver-class>org.postgresql.Driver</driver-class>
  <driver>postgresql-9.4-1200-jdbc41.jar</driver>
  <security>
    <user-name>postgres</user-name>
    <password>postgres</password>
  </security>
  <validation>
    :
</datasource>
```

Step Three

Next you only need to address the configured data source with

```
<jta-data-source>java:/PostgresTraccar</jta-data-source>
<jta-data-source>${datasource.jndi}</jta-data-source>
```

with the latter configured in the Maven repository context.

The main idea of creating the JEETS EAR file is to demo the interaction with the application server. To keep the overview you should go through the startup and or deployment log

```
WFLYSRV0027: Starting deployment of "jeets-jee-app.ear" ..
WFLYSRV0027: Starting deployment of "activemq-rar.rar" ..
WFLYSRV0027: Starting deployment of "postgresql-9.4-1200-jdbc41.jar" ..
WFLYSRV0207: Starting subdeployment (runtime-name: "jeets-jee-ejb.jar")
WFLYSRV0207: Starting subdeployment (runtime-name: "jeets-jee-web.war")
```

... and verify the processing of each resource

```
WFLYWELD0003: Processing weld deployment jeets-jee-app.ear
  HHH000204: Processing PersistenceUnitInfo [name: jeets-pu-traccar-jee ...]
WFLYWELD0003: Processing weld deployment jeets-jee-web.war
WFLYWELD0003: Processing weld deployment jeets-jee-ejb.jar
WFLYEJB0473 : JNDI bindings for session bean named 'ApplicationBean'
              in deployment unit 'subdeployment "jeets-jee-ejb.jar"
              of deployment "jeets-jee-app.ear"' ...
WFLYJPA0010 : Starting Persistence Unit (phase 2 of 2) Service
              'jeets-jee-app.ear#jeets-pu-traccar-jee'
```

With several different technical resources you should be able to pick the ones for your system and change the configuration as needed.

Nevertheless the provided JEETS EAR file does not interact with the database to keep the seed component simple. First you should carefully consider, if your application really needs database access. Or you might want to use a different PU, like the one for GTFS to lookup schedules etc.

For a complete example that you can apply to the EAR as a template you should run the example [60]. This example also shows you how to interact via REST, which you can use to replace JSF – to meet the Web designer's skill set. In the example you can also identify how to connect the code to the database in the `JPAComponentProducer` class:

```
@Produces
@PersistenceContext(unitName = "jeets-pu-traccar-jpa")
private EntityManager em;
```

Okay, now that we've configured WildFly according to the resources in our JEETS EAR, i.e. JEETS App, we are still missing the Camel routing of device messages through the application.

We'll do exactly that in the following sections.

19.2   PUTTING IT ALL TOGETHER

This is your final practical JEETS exam to roll out your own GTS. You should have compiled, packaged and installed the complete JEETS repository at this point. Now you can apply the created project files, i.e. Java applications, to compose a tracking system with the following components and layers:

19.2.1   Client Software (Simulation)

The client side is complementary to the GTS and we need it to send messages to process through the GTS. We are prepared to do this with the JEETS components `jeets-tracker`, `jeets-player` or a more complex source coded from `my-jeets-client`.

Let's revisit our GTFS factory with data from the HVV agency in Hamburg. You can setup your `my-jeets-client` to replay a single transit vehicle or you can roll out one factory for all vehicles at one time! To achieve this you can use a GTFS data source or apply to access the REST API called GeoFox and retrieve live traffic.

Usually transit vehicles are defined with line keys for different vehicle types like `L:` for the line, `B` for the agency and `U`, `S`, `B` for the vehicle type.

For our setup we'll instantiate `my-jeets-client` two times to simulate the Subway `U1` from Fuhlsbüttel to Farmsen and in the opposite direction. In order not to depend on individual line keys at different daytimes we'll call the devices `U1.FuFa` and `U1.FaFu`. The client software can then be launched twice with slightly different command lines in opposing directions:

```
my-jeets-client "U1" "Fuhlsb\"uttel" "Farmsen" "U1.FuFa"
my-jeets-client "U1" "Farmsen" "Fuhlsb\"uttel" "U1.FaFu"
```

Prepare your client software or hardware to send messages to an `IP:port`
If you are using the `jeets-player` you should see the messages
`'U1.FaFu' sending 2 Positions to 'localhost:5200'`
`at 1524314269511 (4 msgs queued)`
When there is no receiver on server side you can stop the client software
and just leave the command line waiting for the RETURN key.
If you (implicitly) apply the `jeets-player` you can keep it running
as it will buffer the messages with their original GPS fixtimes
which is convenient especially at development time.
And it simulates the hardware behavior of a tracker
loosing contact to the server for a while.

19.2.2 Client Hardware

Of course you can also enrich the scenario with real tracking, choose your protocol from more than 100 Traccar protocols and adjust your port to accept messages on server side.

19.2.3 JSE Server Middleware

We have learned that the layer between client and server is called middleware and mediates between the processing cores. Here we have prepared the JEETS components `jeets-dcs`, `jeets-etl` and `jeets-dcs-amq`.

The most simple use case is to apply the `jeets-dcs-amq` and direct the messages from port to message queue, where they can be buffered until a consumer picks them up.

You might also choose to apply a `GeoBasedRoute` from the `jeets-geo-router` along the path to filter messages for the Hamburg or your cities area. From performance perspective you should keep in mind that Java objects are serialized und unserialized when traveling from one JVM to another. On the other hand the convenience of adding and removing logical steps along the route outweighs these milliseconds.

Launch the message broker
```
.../apache-activemq-5.15.0/bin/activemq start
```
Launch DCS to MQ:
```
java -jar jeets-dcs-amq-0.1-jar-with-dependencies.jar
```
Check output:
```
Netty consumer bound to: localhost:5200
route1 started and consuming from: tcp://localhost:5200
```
If the client software is not running, start it now.

Check client output:
```
'pb.device' sending 1 Positions to 'localhost:5200' ...
```
Check DCS to MQ output:
```
[Camel Thread2 - NettyServerTCPWorker]
INFO route1 - device from Netty: uniqueid: "pb.device"
```
Check frontend `localhost:8161/admin/queues.jsp`
for enqueued messages.

19.2.4 JEE Server Application Scenario

Since there is no consumer running yet the messages should not be dequeued. Now we are ready to finally start the consumer, being our JEE server application. Start WildFly and check startup log with installed JEETS ear compiled from .../jeets-server-jee/jeets-jee-app as described on page 241:

```
Camel CDI is starting Camel context [camel-activemq-context]
Apache Camel 2.20.2 (CamelContext: camel-activemq-context) is starting
Camel context starting: camel-activemq-context
Bound camel naming object: java:jboss/camel/context/camel-activemq-context
Successfully connected to tcp://localhost:61616
Route: route1 started and consuming from: activemq://queue:hvv.in
Total 1 routes, of which 1 are started
Apache Camel 2.20.2 (CamelContext: camel-activemq-context)
        started in 11.200 seconds
```

You can see that Camel is starting the [camel-activemq-context] via CDI. So we have to look in the /web subproject to find the route Builder class:

```
@ApplicationScoped
@ContextName("camel-activemq-context")
public class ActiveMQRouteBuilder extends RouteBuilder {
```

This class is the entry to our EAR construct and directs the Device messages to the ApplicationBean EJB with our prepared method processMessage:

```
from("activemq:queue:hvv.in")
.to("ejb:java:global/jeets-jee-app/jeets-jee-ejb/
                    ApplicationBean?method=processMessage")
```

As you can see Camel routes from the external activemq: to the WildFly internal ejb: resource. Now all you have to do is add some log statements before, inside and after the method invocation to observe the processing:

```
[route2] (Camel (camel-activemq-context) thread #3 - JmsConsumer[hvv.in])
   message: Device [id=0,  ...] to application bean
   application: process message from device 'pb.device'
   Send message Device [id=0, ...] to another bean
```

... and we're done!! From here on you can hook up to your existing JEE application with the major actors being the EJBs accessed by CDI Controllers visualized with JSF.

Now you can hook up your application's beans and controllers. If you have a busy developer team working on your enterprise's core application you might not want to bother them with Camel and they can use the device messages inside their plain Java code and do whatever they want.

If you don't have a JEE application we'll go through a general purpose application in pseudo code in the next section.

19.3   APPLICATION SCENARIOS

With the JEETS repository you have a toolset to create a GTS or a 'Real World Live Application' from JEETS components. The GTFS factory is a great helper to get started very fast. You can simulate traffic with trains, subways, ferries and buses in a single city. As an example you could create a scenario from the Hamburg's GTFS data as described in chapter 11. In section 19.2.1 we have setup `my-jeets-client` to simulate two subways U1 in opposite direction. You can easily raise the traffic in `your-jeets-client` along your development time to simulate a hundred buses driving all over the city.

By starting out with two subways you can already develop a `inDistance` trigger by evaluating each incoming message against all participating vehicles. As soon as the two subways are within a distance of 100 meters they are informed by the application controller, they are introduced to each other and can directly exchange information. Inside the software scenario the vehicles can be provided with some connection to each other to simulate external sensors inside the vehicle.

We should keep the big picture of self driving cars in mind, which requires much more than GTS tracking. Besides developing server software to coordinate cars each car is developed to deal with the incoming information and actually derive a decision for the next maneuver.

The client being a vehicle in real traffic can also be modeled and act inside the server scenario. These vehicles can be modeled as an `abstract vehicle class` and inherited by a `Car` or a `Bus` etc. In the JEE environment we can create a `@Stateful` bean for each vehicle for the application time (or whatever scope fits your needs).

19.3.1   Camel Routing

In the applications main Camel route you could parse the vehicle's unique ID and direct it to a stateful EJB. If the vehicle's EJB already exists the EJB is already instantiated, if not it will be created. A `ScenarioController` holds a collection of participating vehicles that can be continuously traversed to determine distances etc. This pseudo code expresses the idea:

```
from("activemq:queue:hvv.in")
.to("...ApplicationBean?method=processMessage") // i.e. VehicleController

.log("Send (modified!) message ${body} to another bean");
// isRegistered? DB lookup !
.choice()
   .when(vehicleType == 'Bus'"))
      .to( bus )
       .when(vehicle == 'AH234:233'"))
       .to( Vehicle implementation Bean ! with environment sensors)
        ...
   .otherwise()
      .log( unknown vehicle - create one? );
```

First the new message is sent to the controller of the scenario (if needed) to be 'in control' over all participating vehicles and latest messages. Then the message is sent to the (virtual) vehicle to adjust its location (if needed). We have discussed different SDC aspects of a server software and by sending the message to a controller it can also aggregate, package and distribute (map link) data to an ftp server and have each vehicle pick up traffic information as needed.

You can also access and sync with other traffic (crowd) sources and create a living city (map) with traffic management, monitoring, control as an information system to switch traffic lights according to traffic!

19.3.2 Application Core

Enterprise Java beans are the main actors and working horses of an JEE environment. Since fleet management is a major scenario in traditional GPS tracking systems we have created an EJB to be instantiated at deployment time to keep track of everything going on until the runtime ends somehow.

```
@Singleton
@Startup
public class ApplicationBean /* implements MyApplication */ {
```

By applying different scopes you could also instantiate a complete fleet with stateful beans once any user of a transport company signs in to the system.

To keep it simple we will only implement the method processMessage receiving messages to start out a very basic controller:

```
public void processMessage(Device devMsg) {
    System.out.println(
        "Application: process msg#" + ++messageCount
            + " from device '" + devMsg.getUniqueid() + "'");
    lastMessage = devMsg;

    // TODO define timeout (ie lastMessage) for each vehicle
    //      and remove from list.

    if ( !vehicles.containsKey( devMsg.getUniqueid())) {
        // create new entry for device and use original Device
        // as container for subsequent messages
        System.out.println("add " + devMsg.getUniqueid() +
            " to collection ...");
        vehicles.put(devMsg.getUniqueid(), devMsg);
    } else {
        System.out.println("add "
            + devMsg.getPositions().size() + " Positions and "
            + devMsg.getEvents().size() + " Events "
            + "to collection " + devMsg.getUniqueid() + "...");
        Device dev = vehicles.get( devMsg.getUniqueid() );
        // add Positions and Events to existing vehicle
        List<Position> allPositions = dev.getPositions();
        allPositions.addAll(devMsg.getPositions());
        dev.setPositions(allPositions);
```

```
        dev.setLastupdate(devMsg.getLastupdate());
    }
}
```

This implementation counts all messages arriving in the application and creates a new vehicle for a messages unique ID or it adds the message if the vehicle already exists. For simplicity each vehicle is represented by a `Device` object and for a new vehicle we simply use the incoming `Device` message, which is the reward of a long and thorough design process. The collection of vehicles is also kept very basic with

```
private Map<String, Device> vehicles = new HashMap<>();
```

With this construct we have created a structure to distribute all incoming messages into a scenario. To inspect the scenario for number of all vehicles and messages and each vehicle with its messages we simply add some getters:

```
public List<Device> getVehicles() { ... }
public Device getVehicle( String uniqueId ) { ... }
public int getMessageCount() { ... }
public Device getLastMessage() { ... }
```

... and we can head on to create a frontend where we can actually retrieve and observe the scenario in a browser ...

At this point each tracking message is propagated to the application scenario and the application developers can forget about the message source and head on to JEE development with an AS.

You will find the complete application (seed) in the JEETS repository and model it to fit your needs. The `ApplicationController` is an application scoped managed bean in the `/web` layer to access the `/ejb` layer, clear separation of layers and project artifacts for different skill sets.

Once you have created your controller you can start adding methods, getters and setters to serve your application. The application frontend can then be modeled with JSF and the web designer only needs to add expressions to access the controller:

```
<h:outputText value="update page " />
<h:commandButton value="Display Server Status" style="width:100%" />
<h:outputText value="last update: #{appcontroller.getCurrentTime()}" />
----------------------------------
```

And there you are in your JEE application where you apply your JEE skills to process the incoming messages!

19.4   JEETS OUTLOOK

Once you have the JEETS application up and running you are a qualified
JEETS developer and should be able to model your own application. This
book has demonstrated how to define your tracking messages towards a system
model, which is materialized in the persistence unit. When the message reaches
the server port it will be transformed and propagated to the system, to your
system.

One last time it should be highlighted that we have modeled the complete
route from tracking client to DCS, ETL, geobased routing and then to a large
JEE application. The higher level modeling actually takes place in the JEETS
repository to become the actual Software Model.

In order to keep the repository together as one system, developers have to
add test cases with each implementation. The project manager can focus on
the development of the repository itself by adding Integration Testing over all
software modules.

Please check the JEETS repository at github for the latest developments.
The roadmap requires the modeling of a complete message route from client
to server and possibly back to the client. A complete system integration test
should test each Maven project individually, then combined projects, like DCS.
After that you can create a client with a tracker, a player, a factory etc. and
run it inside your test environment. The server can be setup with a clean in-
memory database, create the model with JPA on the fly, accept the incoming
messages and direct them to the modules to be tested. In another development
cycle the system test can be tested for performance...

Then after you have modeled the complete transmission of a GPS live
track from client to server you can add Arquillian to launch WildFly, add a
message broker for testing and you can supply a real database connection to
compare in memory performance with the installed RDBMS. Each external
resource should be supplied with a Maven profile to provide control over each
dedicated testing environment.

As you can see there is still a lot of work to ensure a stable GTS like
Traccar and we will see where it's heading at `jeets.org`. The creation of
various integration environments is not related to GPS tracking, just like
JEE programming with different subsystems. Therefore we will conclude the
book at this point and head on to a shared repository for discussions and
improvements.

See you at `jeets.org` !

Development Environment

CONTENTS

This book was written from the developer seat. coding, running and writing section by section. While being introducing to JEE components like Netty the reader is expected to go through their introduction, quick starts and tutorials as appropriate to follow the book development.

For deeper research many components are associated with at least one complete book and many are actually written by the component owners of Netty, JPA, Hibernate, Camel, ActiveMQ, etc.

This appendix provides installation instructions in order of appearance in the book. You can choose to go through the installation and testing instructions all at one time. Or you will be pointed to the steps required at the beginning of a chapter.

A.1   PREREQUISITES

This book is about Java EE and the reader should be familiar with the typical build process. The project provides a single GIT repository to download, compile and test the code with Maven. This environment should run without IDE dependencies and can be added to existing continuous development and integration procedures.
Please install the following components now:

Java 1.8 JDK 1.8...

Apache Maven 3.3...

Git version control system
If you don't want to publish any code you only need git for downloading. SourceTree is a free GIT frontend by Atlassian you can use.

A.2 IDE

This book is for developers and does not cover the deployment to production systems. For example you might choose to run a server instance of the Traccar GTS to track yourself. In this case you should refer to the Traccar project home to pick a binary distribution for your hosting server OS.

In order to follow the books analysis you should download the Traccar sources and import them in your IDE – as explained in the next appendix. During the writing of the book

Eclipse Java EE: Neon Release (4.6.0)

was used. Make sure to download the Java EE IDE with

JBoss Tools

EGit

Web Tools Platform

Data Tools Platform

Maven m2e

It is recommended to create a new workspace for all sources of the book.

Install Traccar Sources

CONTENTS

This appendix *does not* describe how to install Traccar for permanent tracking on a server with a fixed IP. You should refer to the Traccar project home to pick a binary distribution for your server OS. If you have a chance to deploy Traccar on a server follow the instructions at `traccar.org` to set it up and start collecting tracks!

For the book's purposes we will download the Traccar source code and import it to Eclipse. From Eclipse we can start, debug and stop the system any time. For 24/7 tracking this solution is not recommended, since you need to keep the development environment running when you leave home. The energy management has to be turned off, you may need to change the dynamic IP every day and open the tracking ports through the firewall. . .

Please keep in mind that this book is looking at the Traccar system as a preexisting production system and should be completely installed 'out of the box'. To keep up with Traccar's high pace the system should remain updatable via github while you should keep track of your changes locally.

B.1   SOURCE INSTALLATION

The complete code of the Traccar Tracking system can be browsed at

```
github.com/tananaev/traccar
github.com/tananaev/traccar-web
```

where the first URL points to the Traccar server and the second to the Web application code. Note that the latter is a subproject of the server and you have to apply git's `--recursive` flag if you plan to use the Web frontend – and you should. Once you have both projects installed you can update them independently. The Web application is a pure Javascript project without a need to compile.

As a result you should get the main folder including the subfolder `traccar-web` in your target directory. To download the Traccar server[1] clone the repo to your workstation with

```
git clone https://github.com/tananaev/traccar.git
```

Change to your local Traccar `server` directory with `pom.xml` and run Maven

```
mvn clean install
```

If all goes well Maven will clean, compile, test and install two jars in the target folder. All external dependencies will be downloaded to your local repo upon first execution. As a result you should get two jar files:

```
tracker-server.jar
tracker-server-jar-with-dependencies.jar
```

Now you can run the server with the preconfigured h2 database and watch how the database structure is created:

```
java -jar target\tracker-server.jar setup\traccar.xml
```

and you should see some output like

```
[main] INFO com.zaxxer.hikari.HikariDataSource - HikariPool-1 - Started.
INFO liquibase: Clearing database change log checksums
INFO liquibase: Successfully acquired change log lock
INFO liquibase: Successfully released change log lock
INFO liquibase: Successfully acquired change log lock
INFO liquibase: Reading from PUBLIC.DATABASECHANGELOG
INFO liquibase: Reading from PUBLIC.DATABASECHANGELOG
INFO liquibase: Successfully released change log lock
```

Next you can kill the process with `Ctrl+c` and open the log file specified in xml to find something like

```
INFO: Operating system name: Windows 10 v10.0 architecture: amd64
INFO: Java runtime name: Java HotSpot(TM) 64-Bit Server VM vendor:
                         Oracle Corporation version: 25.101-b13
INFO: Memory limit heap: 1797mb non-heap: 0mb
INFO: Character encoding: Cp1252 charset: windows-1252
INFO: Version: null
INFO: Query not provided: database.selectAttributeAliases
INFO: Starting server...
INFO: Shutting down server...
```

and as you can see in the last line the server is being shut down by Traccar and not simply killed to leave connections open etc. Good job!

You have now successfully downloaded, compiled, tested and run the Traccar server! Keep the command line window open as we will run it again soon. Before importing the project into your IDE it is highly recommended to install PostgreSQL and configure it for Traccar as described in the next section.

---

[1] At the time of writing v3.14

Note that we have installed Traccar without activating the frontend

```
<!-- SERVER CONFIG -->
<entry key='web.enable'>false</entry>
```

since we first want to make sure the server is running without a frontend. Make sure the logging is enabled and remember the target folder to look for the log files. Before we install PostgreSQL in the next section check in the DATABASE CONFIG section to see if the h2 database is active. Its a database for development purposes only and needs no installation.

Of course tracking is much more fun with a browser frontend and a map. Therefore you should figure out how to enable it ;) You can go through the same procedure as above and on server start you should see entries with 'jetty'

```
[main] INFO org.eclipse.jetty.
```

Jetty [50] is a popular Java HTTP (Web) server and Java Servlet container and is often used for machine to machine communications.

B.2   POSTGRESQL INSTALLATION

First make sure your server is running as described in the previous section. For JEETS we will use PostgreSQL for our Database Managment system (DBMS) as the preferred database for geospatial processing. Please install the latest Postgres server (postgresql-9.5.4...) in your environment and make sure to add the pgAdmin frontend. After installing Postgres we need to modify the xml file[2]:

```
<entry key='database.driver'>org.postgresql.Driver</entry>
<entry key='database.url'>
            jdbc:postgresql://127.0.0.1:5432/traccar</entry>
<entry key='database.user'>postgres</entry>
<entry key='database.password'>postgres</entry>
```

where traccar will point to the Traccar database model. Password and User *can* be identical to the main Postgres credentials, which is a common setup for development. After installation this database doesn't exist yet, so we have to create it:

1. Open pgAdmin, connect to the installed postgres instance

2. Right click on the databases and select new database

3. Enter traccar as the database name and click OK

4. Go back to the command line to start the Traccar server

---

[2]see www.traccar.org/postgresql

This time you should see additional entries like

```
INFO 9/14/17 3:34 PM: liquibase: Reading from public.databasechangelog
INFO 9/14/17 3:34 PM: liquibase: ./schema/changelog-master.xml:
  changelog-3.3::changelog-3.3::author: Table users created
            :
INFO 9/14/17 3:34 PM: liquibase: ./schema/changelog-master.xml:
  changelog-3.15::changelog-3.15::author: ChangeSet
  changelog-3.15::changelog-3.15::author ran successfully in 54ms
INFO 9/14/17 3:34 PM: liquibase: Successfully released change log lock
```

which is a table creation script. If you refresh the database in pgAdmin you should find the new Traccar database with its tables! Stop and start the server again to make sure the database is not created again to override a possibly existing and filled database.

Traccar is now ready to receive and collect tracking messages! The sending and receiving of the first message is described from Section 3.3 on. Before going back there we want to import Traccar into Eclipse where we can inspect the incoming message.

Important Note!
In the course of the book you should always apply the original Traccar updates to the software *and* database! The persistence unit created in the book can also be used to create tables. On the other hand the book only maintains the PU with respect to the JEETS source code. The PU does create the complete database, but not all constraints have been named yet and the Traccar update mechanism would run into problems[3]

```
CONSTRAINT fk_user_device_geofence_deviceid FOREIGN KEY (deviceid)
CONSTRAINT fkc5ce2s99d3p0dtopuc5hrckc      FOREIGN KEY (deviceid)
```

On the other hand you *can* apply the persistence unit on the original database *after* the modifications for a new version have been applied. For a longer term the JEETS repository should include a test case to create a database schema with Traccar, run it with the PU tests and vice versa – for different version states.

B.3   TRACCAR UPDATE

Traccar is under constant development and it would be a pity not to apply bugfixes and new features to our 'production system'. Naturally the releases after the publication of this book can not be described here and you should visit **jeets.org** and of course the repo version to check the latest status.

The book approach is to prototype tracking components along with (or against) the Traccar system. This way the Traccer frontend can still be used to

---

[3]Constraints can also be defined in the PU. Due to time restrictions maintenance is pending. Please check latest version at **www.jeets.org**

administer people and devices You should always be aware of this constellation when you add code. The JEETS repo does not push any changes to the Traccar repository.

Before you pull Traccar updates make sure to create a backup archive of the current status, to commit your changes in advance. Then you can run `mvn clean install` and go through the installation procedure described earlier. Then you can reinsert your `*.xml` files for customization and point the database driver to your Postgres instance.

```
<entry key='database.url'>
    jdbc:postgresql://127.0.0.1:5432/traccar</entry>
```

The adoption of the persistence unit is described in Chapter 6 with its intricacies and should be well understood, if you plan to develop your own architecture.

B.4   TRACCAR IMPORT TO ECLIPSE

After installing Eclipse and the Traccar sources on your PC select

```
File > Import > Existing Maven Project > <traccar root>
```

and as an exercise you can run Maven from Eclipse m2e to get the same results as earlier. To launch Traccar (in debug mode) create a 'run configuration' with

```
            Project: traccar
         Main class: org.traccar.Main
  Program arguments: path/to/traccar3/server/setup/jeets.xml
```

Now you are ready to receive the first message.
Please continue reading in Section 3.3.

Install JeeTS Sources

CONTENTS

All JEETS components are bundled in a single GIT repository at github. To ensure their functionality and compatibility unit tests are implemented for every single component to ensure stand-alone usage. The built process will start with the independent components and add dependencies of higher level components. If you modify the sources you should make sure that all tests (starting with the modified module) return a green bar to maintain integrity.

C.1   BUILDING THE JeeTS REPOSITORY

Clone the JEETS repository to your PC:

```
git clone https://github.com/kbeigl/jeets.git
```

Change to your local Traccar directory (with `pom.xml`) and run Maven:

```
mvn clean        // validate Maven installation
mvn compile      // compile *.java sources
mvn test         // tests without additional requirements
mvn test -P...   // tests with Postgres database and Traccar user
mvn test -P...   // tests with running ActiveMQ Service
mvn test -P...   // tests with running Wildfly Server
mvn install      // all of the above ???
```

If all goes well Maven will clean, compile, test and create a target folder for each artifact. All external dependencies will be downloaded to your local repo upon first execution.

With modern tooling you can implement anything in test cases. The JEETS repository should be perceived as a number of projects which can be tested stand alone. Other projects depending on the platonic modules can run through other test setups. Therefore the repository is created in the order of low level Modules, like the Hibernate PU. Each module runs through basic tests, if you do not specify a Maven profile.

C.2   ECLIPSE SOURCE IMPORT

You *can* import the complete JEETS repository into Eclipse. Anyhow it is more practical to build the JEETS repository from command line with Maven.

After you have successfully compiled you should open Eclipse with a new workspace for the JEETS projects. Then you choose

```
File > Import ...
       > Existing Maven Projects
       > Root Directory <\pathTo\repo.jeets\>
```

Now you should see the complete JEETS structure and you can apply the checkboxes to the project/s you want to work with. Note that it is advised (sometimes mandatory) to use Maven on the command line even if you are working with Eclipse! Eclipse m2e will retrieve the precompiled and installed depend projects from your local Maven repository in the .m2 folder.

C.3   PROTOBUFFER COMPILER

The `traccar.proto` file is the reference file for the JEETS TCP protocol and should be reversioned after each change. It resides in the folder

```
.../repo.jeets/jeets-models/jeets-protocols/protobuffers
```

From this file the Java class

```
.../jeets-protocols/src/main/java/org/jeets/protocol/Traccar.java
```

is generated with the protobuffer compiler `protoc`. Then this Java file becomes part of the JEETS repository and developers can compile against it.

In order to understand this process you should practically go through the Java Protobuffers Tutorial at

```
developers.google.com/protocol-buffers/docs/javatutorial
```

After you have concluded the tutorial you should create a backup of the original `Traccar.java` class. Then you can use the `protoc` compiler on the `traccar.proto` file to create the `Traccar.java` file.

For your own protocols you should also start out with a `*.proto` file and create message formats according to your system model.

C.4   ACTIVEMQ INSTALLATION AND TESTS

Note that ActiveMQ not only requires Maven to execute the examples and ANT should be installed to run the `swissarmy` example to validate the installation and functionality.

You can pick your installation according to your OS at

activemq.apache.org/version-5-getting-started.html

where you will find the Installation Procedure.
This is a rough guideline that might help you

```
extract ANT C:\prox\apache-ant-1.10.1
set ANT_HOME and path %ANT_HOME%\bin
cd ANT_HOME
ant -f fetch.xml -Ddest=system

extract apache-activemq-5.15.0-bin.zip
bin\activemq start
administrative interface  http://127.0.0.1:8161/admin/
Login: admin  Passwort: admin

start the broker with the demos included:
bin\activemq.bat start xbean:examples/conf/activemq-demo.xml

...\apache-activemq-5.15.0\examples\openwire\swissarmy .. success
```

Install GTFS Sources

CONTENTS

D.1   DISCLAIMER

Please note that GTFS data is publicly available for download. This does not imply that you can do with it whatever you choose! If you plan to publish a website with GTFS data you should carefully study the legal issues and maybe even get in touch with the agency providing the data. Once you are in contact with the agency you might even get access to their REST API providing live vehicle information!

For testing and development of JEE components you can download any GTFS dataset and create traffic on your development PC.

D.2   FINDING YOUR GTFS DATASET

With GFTS google has provided a global specification for public transit schedules and with it a great number of applications have emerged. For our purposes we don't want to get lost in the world of GTFS and go straight to creating a GTFS database with Postgres and – very important – with PostGIS for geographical indexes.

The JEE code for the transit factory was developed with real GTFS data for the city of Hamburg in Germany. Due to the standard format the code should also work for your City and it is worth while to search for its dataset. Just by searching for 'GTFS' you will find many overview sites linking to the actual sources for a city.

Hamburg, like more and more cities, is providing a 'Transparency Portal' with a large number of 'open data' sets including GTFS. At transparenz.hamburg.de you simply search for 'GTFS' to find something like

HVV-Fahrplandaten (GTFS) November 2017 bis Dezember 2017

```
Veroeffentlichende Stelle: Hamburger Verkehrsverbund GmbH
Veroeffentlichungsdatum:    03.11.2017

Der Hamburger Verkehrsverbund (HVV) stellt hier die Fahrplandaten
des Verbundgebietes monatlich im GTFS-Format zur Verfuegung.
Die Daten beinhalten Informationen zu Linien, Haltestellen,
Abfahrtszeiten usw.

Name:   Upload: HVV_Rohdaten_GTFS_Fpl_20171103
Format: ZIP (size: 25.6 MB 150 MB unzipped!)
```

Please search for GTFS data in the city you know best or better the city you
live in and load it to your PC.

D.3   GTFS INSTALLER

Note that the following instructions support you to load GTFS to Postgres /
PostGIS. You should have the Traccar tracking system installed with Postgres,
but if you have chosen a different DBMS the next steps are only guidelines
that you need to modify for your system accordingly.

In the downloaded the dataset you should find these basic tables, i.e. csv
files:

agency.txt	routes.txt	stops.txt
calendar.txt	shapes.txt	transfers.txt
calendar_dates.txt	stop_times.txt	trips.txt

Like many other agencies the Hamburg dataset comes without fares and if you
plan a scheduling service for public users you should ask for this additional
data or for the REST API.

With the dataset we will setup the production chain to import data into
Postgres. Although there are many loading tools available in Java we have
chosen to use the loader from [28] which is written in Python and makes a
stable and maintained impression.

GTFS Database [29]

"The gtfsdb project's focus is on making GTFS data available in a programmatic context for software developers. The need for the gtfsdb project
comes from the fact that a lot of developers start out a GTFS-related effort
by first building some amount of code to read GTFS data (whether that's
an in-memory loader, a database loader, etc...); GTFSDB can hopefully
reduce the need for such drudgery, and give developers a starting point
beyond the first step of dealing with GTFS in .csv file format."

This free standard tool also implies its 'raw' data model which is sufficient for our needs. If you plan to create a GTFS software you might add some tools in the chain to do some cleaning up to increase the performance:

Remove redundant data

Switch from String to Integer keys

Reduce number of shapes with PostGIS Douglas Peucker Algorithm

etc.

D.4 INSTALL GTFSDB

You will find the software with instructions at

`github.com/OpenTransitTools/gtfsdb`

Here is a brief description of the steps you need to take:

Download and install python 2.7..

Download and install setuptools-36.5.0

Download and install psycopg2-2.6.2.win-amd64-py2.7-pg9.5.3-release.exe

Command: pip install zc.buildout

Command: buildout install prod postgresql

After the software is installed you should go to the psycopg2 installation

`...\Python27\Lib\site-packages\psycopg2`

look for the `__init__.py` file and enter the credentials for your GTFS database

```
def connect(dsn=None,
    database='HVV-20171006', user='postgres', password='postgres',
    host='localhost', port=5432,
    connection_factory=None, cursor_factory=None, async=False,
    **kwargs):
```

As a measure of precaution we have named the database after the download file

```
Upload__HVV_Rohdaten_GTFS_Fpl_20171006.zip
database: HVV-20171006
```

but for the longer term you shouldn't include the `yyyymmdd` date in the name and use the same name for subsequent deliveries. Since you usually request transit information for 'today' the data will 'run out' at a certain day and won't return any information. Then you would need to go through the re/installation process again.

Before starting the import you need to create the database in Postgres. This can easily be achieved with the PG admin tool. Then you must connect to the new database and enable PostGIS by entering the SQL command.

`CREATE EXTENSION postgis;`

Another thing you should know is that the import process consumes a lot of CPU power and RAM and it is best you should not work on the PC during the import. The above mentioned dataset with about 150 MB uncompressed data takes about 30 minutes and up to six GB of RAM. The city of Hamburg has about 24.000 stops with about 1.4 million stop times and 2.5 million shapes.

Now you can launch the import (with geospatial processing) with the (single!) command line

```
bin\gtfsdb-load --database_url postgresql://postgres@localhost:5432
   --is_geospatial ..\GTFS\Upload__HVV_Rohdaten_GTFS_Fpl_20171006.zip
```

. . . and you are ready to run the public transit factory described in Chapter 11.

Install WildFly with Camel

CONTENTS

This is only a brief description of how to install WildFly with Camel 'out of the box'. This minimum installation serves the architect to add Camel routes into and out of the application server environment. From there on you have to take over and configure everything according to your secured production system. By setting up Camel endpoints inside your existing JEE application you will be able to direct tracking messages to it and add them to your business logic.

E.1   INSTALL WILDFLY

Basically a WildFly server is a self contained folder structure and the installation can be automated with build tools. You can automate the following steps, if you intend to include the complete installation in your repository.

1. Download WildFly at `wildfly.org/downloads/`
   i.e. `wildfly-11.0.0.Final` (`.zip` `.tar.gz`)
   add `JBOSS_HOME` to your OS

2. Go to `JBOSS_HOME`
   run `bin/standalone` (`.sh`|`.bat`)
   check `localhost:8080`

3. Run `add-user.sh`
   remember user and password
   login at `localhost:9090`

You will find all the details at
    docs.jboss.org/author/display/WFLY/Getting+Started+Guide

E.2   INSTALL CAMEL SUBSYSTEM

Please visit [57]:

The WildFly Camel User Guide

> "... provides Camel integration with the WildFly Application server.
> The WildFly-Camel subsystem allows you to add Camel Routes as part of the WildFly configuration. Routes can be deployed as part of JavaEE applications. JavaEE components can access the Camel Core API and various Camel component APIs.
> Your Enterprise integration Solution can be architected as a combination of JavaEE and Camel functionality."

Download the release for your WildFly version at

    github.com/wildfly-extras/wildfly-camel/releases

which you can find in the WildFly compatibility matrix.
For this book we have used

    wildfly-camel-patch-5.1.0.tar.gz   294 MB

to provide `Camel-2.20.2` integration with `WildFly-11.0.0` If you look at the content of the archive you can see part of the folder structure of the WildFly installation.

1. Download WildFly-Camel

2. Move archive to `JBOSS_HOME`

3. Go to `JBOSS_HOME`
   Run `bin/standalone` (.sh|.bat) `-c standalone-camel.xml`

4. Login to the management console at `localhost:9090` and look for the Camel subsystem in the configuration section

5. If you have configured WildFly in your IDE make sure to apply the `standalone-camel.xml` file for launching.

Now you are ready to deploy your first Camel component on WildFly!

E.2.1 Wildfly-Camel Configuration File

If you already have a WildFly instance running for your existing system you should use your original `standalone.xml` file and merge the Camel stuff into it. If you compare the original `standalone.xml` file provided from the WildFly installation to the `standalone-camel.xml` file you can easily identify the few lines to add to your configuration. In a merge view you can add the missing portions with some mouse clicks.

E.2.2 Hawtio Web Console

The Hawt Web console has been installed with the WildFly-Camel Subsystem and you should be able to access it at `localhost:8080/hawtio/`. You can login with the WildFly application user you have created earlier.

If you choose the JMX tag and open the `jboss.as` node you can actually look into the application server. If you have installed everything described above you should be able to locate the `activemq-rar.rar` and the `hawtio-wildfly-1.5.7.war` among many other components.

Let's connect to another Java resource - the ActiveMQ installation. Earlier we looked at the traversing messages with the ActiveMQ frontend at `localhost:8161`. At `localhost:8161/api/jolokia/` you can retrieve server infos via REST API. If not you should check your `activemq.xml` file and set

```
<managementContext>
    <managementContext createConnector="true"/>
</managementContext>
```

To make life easier we can also use hawt to connect to ActiveMQ. Choose the 'connect' tab and enter the connection parameters:

```
scheme: http
  host: admin:admin@localhost
  port: 8161
  path: api/jolokia
```

Note that you can add user and password to the host URL. Press 'connect to server' to open a new browser tab and explore the ActiveMQ status.

As an alternative setup you can also embed ActiveMQ in WildFly and the component can be configured to work with an embedded or external broker. For Wildfly / EAP container managed connection pools and XA-Transaction support, the ActiveMQ resource adapter can be configured into the container configuration file. WildFly resource adapters subsystem is wired to the ActiveMQ adapter.

Now you are ready to test your first Camel route
in Section 18.5 'WildFly-Camel Quickstart'.

Recommended Readings

[1] traccar.org

[2] brettwooldridge.github.io/HikariCP/

[3] www.liquibase.org

[4] www.eclipse.org/eclipselink/

[5] www.vehicletrackingsystem.in/blog/tag/tk103-protocol-example/

[6] developer.garmin.com/fleet-management/overview

[7] developer.garmin.com/fleet-management/protocol-support/

[8] www.garmin.com/support/pdf/iop-spec.pdf

[9] docs.oracle.com/javase/tutorial/networking/

[10] tools.ietf.org/html/rfc862

[11] www.jcabi.com

[12] project home: netty.io

[13] netty.io/wiki/user-guide-for-4.x.html

[14] netty.io/4.0/xref/io/netty/example/echo/package-summary.html

[15] project home: mina.apache.org

[16] project home: www.postgresql.org

[17] developers.google.com/protocol-buffers

[18] developers.google.com/protocol-buffers/docs/javatutorial

[19] Kristof Beiglböck
Programming GPS and OpenStreetMap Applications with Java
CRC Press, Inc. Boca Raton, FL, USA, 2011.

[20] project home: hibernate.org

[21] hibernate.org/orm/

[22] hibernate.org/tools/

[23] hibernate.org/ogm/

[24] maven.apache.org/guides/introduction/introduction-to-archetypes.html

[25] en.wikipedia.org/wiki/General˙Transit˙Feed˙Specification

[26] developers.google.com/transit/gtfs/reference/

[27] The Definitive Guide to GTFS by Quentin Zervaas

[28] github.com/OpenTransitTools

[29] github.com/OpenTransitTools/gtfsdb

[30] www.eaipatterns.com or enterpriseintegrationpatterns.com

[31] Gregor Hohpe, Bobby Woolf *Enterprise Integration Patterns: Designing, Building, and Deploying Messaging Solutions* Addison-Wesley, 2004

[32] www.enterpriseintegrationpatterns.com/
    ... patterns/messaging/MessageTranslator.html

[33] eaipatterns.com/patterns/messaging/Message.html

[34] project home: camel.apache.org

[35] camel.apache.org/components.html and ../component-list-grouped.html

[36] camel.apache.org/netty4.html

[37] camel.apache.org/protobuf.html

[38] camel.apache.org/type-converter.html

[39] camel.apache.org/using-getin-or-getout-methods-on-exchange.html

[40] camel.apache.org/seda.html

[41] camel.apache.org/spring-example.html

[42] camel.apache.org/jpa.html

[43] camel.apache.org/hibernate.html and /hibernate-example.html

[44] camel.apache.org/content-based-router.html

[45] camel.apache.org/hiding-middleware.html

[46] camel.apache.org/pojo-consuming.html

[47] camel.apache.org/bean-binding.html

[48] camel.apache.org/tutorial-jmsremoting.html

[49] camel.apache.org/jetty.html

[50] project home: eclipse.org/jetty

[51] camel.apache.org/activemq.html

[52] project home: activemq.apache.org

[53] activemq.apache.org/how-do-i-embed-a-broker-inside-a-connection.html

[54] wildfly-extras.github.io/wildfly-camel/#˙getting˙started

[55] wildfly-swarm.io

[56] project home: arquillian.org

[57] github.com/wildfly-extras/wildfly-camel/

[58] github.com/wildfly-extras/wildfly-camel-examples

[59] .../wildfly-camel-examples/tree/master/camel-jms-mdb

[60] .../wildfly-camel-examples/tree/master/camel-jpa

[61] pdfbox.apache.org

[62] salilstock.blogspot.de ...
      .../2016/03/direct-vs-seda-vs-direct-vm-vs-vm-end.html

[63] project home: spring.io

[64] projects.spring.io/spring-framework/#quick-start

[65] wiki.openstreetmap.org/wiki/Slippymaptilenames - Java method
      wiki.openstreetmap.org/wiki/Zoomlevels - tile sizes, map scales etc.

[66] github.com/locationtech/jts

[67] www.locationtech.org/projects/technology.jts

[68] github.com/locationtech/jts/blob/master/MIGRATION.md

[69] JSR 345 – jcp.org/en/jsr/detail?id=345
      download.oracle.com/ ...
      ...otn-pub/jcp/ejb-32-fr-eval-spec/ejb-32-core-fr-spec.pdf
      ...otn-pub/jcp/ejb-32-fr-spec/ejb-32-optional-fr-spec.pdf

[70] en.wikipedia.org/wiki/Java˙Management˙Extensions

[71] camel.apache.org/camel-jmx.html

[72] project home: hawt.io

Index